AF522065

A-Z
THERMODYNAMICS

A-Z
THERMODYNAMICS

Prof. G.K. Bose

CENTRUM PRESS
NEW DELHI-110002 (INDIA)

CENTRUM PRESS
H.O.: 4360/4, Ansari Road, Daryaganj,
New Delhi-110 002 (India)
Ph.: 23278000, 23261597
B.O.: No. 1015, Ist Main Road, BSK IIIrd Stage
IIIrd Phase, IIIrd Block,
Bangalore - 560 085 (India)
Tel.: 080-41723429
Visit us at: www.centrumpress.com

A-Z Thermodynamics

First Edition, 2009
ISBN 978-93-80106-39-7

PRINTED IN INDIA

Printed at Salasar Imaging Systems, Delhi-110035 (India)

Contents

Preface

Thermodynamics often confuses a beginner. And, there are valid reasons for it. Let us, for instant, consider how the fundamental laws of thermodynamics are introduced to an absolute beginner. Many texts introduce the first law by the statement 'When any closed system is taken through a cycle, the net work delivered to the surroundings is proportional to the net heat taken from the surroundings.' And, the second law is introduced by the Kelvin-Plank and the Clausius statements, that are about the limits placed in physically realizing engines whose working fluids operating in cyclic processes. Such an introduction to thermodynamics and teaching of the resulting corollaries, I realized, only made the young people grew indifferent to thermodynamics. Majority of them wrote in their course-teacher evaluation sheets that they found it hard to relate to thermodynamics. They, the students, made one thing very clear to me that, if they were to like the subject thermodynamics, the subject matter should be presented in a simple-and-easy-to-understand style.

Atmospheric thermodynamics is often the first quantitative atmospheric science course encountered by new meteorology majors. This is not surprising, as it serves as the natural bridge between the generic principles of physics and calculus (energy and work, differentiation and integration) acquired in the lower division years and the more specialized knowledge and methods associated with atmospheric dynamics and the analysis of weather systems.

Author

Chapter 1

Thermodynamics

Thermodynamics is a branch of physics which deals with the energy and work of a system. It was born in the 19th century as scientists were first discovering how to build and operate steam engines. Thermodynamics deals only with the large scale response of a system which we can observe and measure in experiments. Small scale gas interactions are described by the kinetic theory of gases. The methods complement each other; some principles are more easily understood in terms of thermodynamics and some principles are more easily explained by kinetic theory.

There are three principal laws of thermodynamics which are described on separate slides. Each law leads to the definition of thermodynamic properties which help us to understand and predict the operation of a physical system. We will present some simple examples of these laws and properties for a variety of physical systems, although we are most interested in thermodynamics in the study of propulsion systems and high speed flows. Fortunately, many of the classical examples of thermodynamics involve gas dynamics. Unfortunately, the numbering system for the three laws of thermodynamics is a bit confusing. We begin with the zeroth law.

The zeroth law of thermodynamics involves some simple definitions of thermodynamic equilibrium. Thermodynamic equilibrium leads to the large scale definition of temperature, as opposed to the small scale definition related to the kinetic energy of the molecules. The first law of thermodynamics relates the various forms of kinetic and potential energy in a system to the work which a system can perform and to the transfer of heat. This law is sometimes taken as the definition of internal energy, and introduces an additional state variable, enthalpy. The first law of thermodynamics allows for many possible states of a system to exist. But experience indicates that only certain states occur. This leads to the second law of thermodynamics and the definition of another state variable called entropy. The second law stipulates that the total entropy of a system plus its environment can not decrease; it can remain constant for a reversible process but must always increase for an irreversible process.

THERMODYNAMIC POTENTIALS

As can be derived from the energy balance equation on a thermodynamic system there exist energetic quantities called thermodynamic potentials, being the quantitative measure of the stored energy in the system. The five most well known potentials are:

Internal energy U

Helmholtz free energy $A = U - TS$

Enthalpy $H = U + PV$

Gibbs free energy $G = U + PV - TS$

Grand potential $\Phi_G = U - TS - \mu N$

Other thermodynamic potentials can be obtained through Legendre transformation. Potentials are used to measure energy changes in systems as they evolve from an initial state to a final state. The potential used depends on the constraints of the system, such as constant temperature or pressure. Internal energy is the internal energy of the system, enthalpy is the internal energy of the system plus the energy related to pressure-volume work, and Helmholtz and Gibbs energy are the energies available in a system to do useful work when the temperature and volume or the pressure and temperature are fixed, respectively.

CLASSICAL THERMODYNAMICS

Classical thermodynamics is the original early 1800s variation of thermodynamics concerned with thermodynamic states, and properties as energy, work, and heat, and with the laws of thermodynamics, all lacking an atomic interpretation. In precursory form, classical thermodynamics derives from chemist Robert Boyle's 1662 postulate that the pressure P of a given quantity of gas varies inversely as its volume V at constant temperature; i.e. in equation form: $PV = k$, a constant. From here, a semblance of a thermo-science began to develop with the construction of the first successful atmospheric steam engines in England by Thomas Savery in 1697 and Thomas Newcomen in 1712. The first and second laws of thermodynamics emerged simultaneously in the 1850s, primarily out of the works of William Rankine, Rudolf Clausius, and William Thomson.

STATISTICAL THERMODYNAMICS

With the development of atomic and molecular theories in the late 1800s and early 1900s, thermodynamics was given a molecular interpretation. This field is called statistical thermodynamics, which can be thought of as a bridge between macroscopic and microscopic properties of systems. Essentially, statistical thermodynamics is an approach to thermodynamics situated upon statistical mechanics, which focuses on the derivation of macroscopic results from first principles. It can be opposed to its historical predecessor phenomenological thermodynamics, which gives scientific descriptions of phenomena with avoidance of microscopic details. The statistical approach is to derive all macroscopic properties (temperature, volume, pressure, energy, entropy, etc.) from the properties of moving constituent particles and the interactions between them (including quantum phenomena). It was found to be very successful and thus is commonly used.

CHEMICAL THERMODYNAMICS

Chemical thermodynamics is the study of the interrelation of heat with chemical reactions or with a physical change of state within the confines of the laws of thermodynamics. On the Equilibrium of Heterogeneous Substances, in which he showed how thermodynamic processes could be graphically analyzed, by studying the energy, entropy, volume, temperature and pressure of the thermodynamic system, in such a manner to determine if a process would occur spontaneously. During the early 20th century, chemists such as Gilbert N. Lewis, Merle Randall, and E.A. Guggenheim began to apply the mathematical methods of Gibbs to the analysis of chemical processes.

THERMODYNAMIC SYSTEMS

An important concept in thermodynamics is the "system". Everything in the universe except the system is known as surroundings. A system is the region of the universe under study. A system is separated from the remainder of the universe by a boundary which may be imaginary or not, but which by convention delimits a finite volume. The possible exchanges of work, heat, or matter between the system and the surroundings take place across this boundary. Boundaries are of four types: fixed, moveable, real, and imaginary.

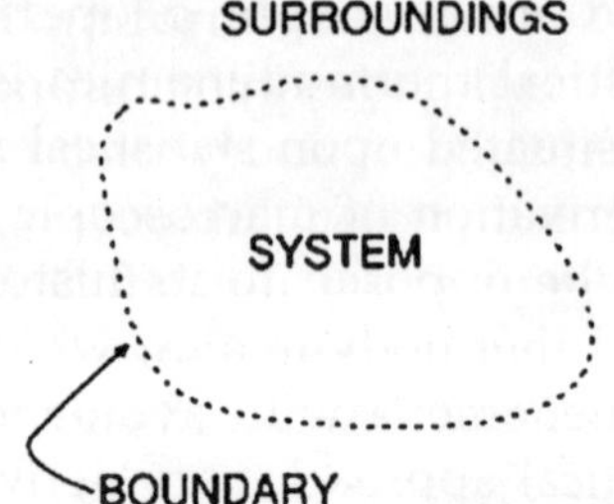

Fig. Thermodynamic Systems

Basically, the "boundary" is simply an imaginary dotted line drawn around a volume of something when there is going to be a change in the internal energy of that something. Anything that passes across the boundary that effects a change in the internal energy of the something needs to be accounted for in the energy balance equation. That something can be the volumetric region surrounding a single atom resonating energy, such as Max Planck defined in 1900; it can be a body of steam or air in a steam engine, such as Sadi Carnot defined in 1824; it can be the body of a tropical cyclone, such as Kerry

Emanuel theorized in 1986 in the field of atmospheric thermodynamics; it could also be just one nuclide (i.e. a system of quarks) as some are theorizing presently in quantum thermodynamics.

For an engine, a fixed boundary means the piston is locked at its position; as such, a constant volume process occurs. In that same engine, a moveable boundary allows the piston to move in and out. For closed systems, boundaries are real while for open system boundaries are often imaginary. There are five dominant classes of systems:

- *Isolated Systems*–matter and energy may not cross the boundary
- *Adiabatic Systems*–heat must not cross the boundary
- *Diathermic Systems*–heat may cross boundary
- *Closed Systems*–matter may not cross the boundary
- *Open Systems*–heat, work, and matter may cross the boundary

As time passes in an isolated system, internal differences in the system tend to even out and pressures and temperatures tend to equalize, as do density differences. A system in which all equalizing processes have gone practically to completion, is considered to be in a state of thermodynamic equilibrium.

In thermodynamic equilibrium, a system's properties are, by definition, unchanging in time. Systems in equilibrium are much simpler and easier to understand than systems which are not in equilibrium. Often, when analysing a thermodynamic process, it can be assumed that each intermediate state in the process is at equilibrium. This will also considerably simplify the situation. Thermodynamic processes which develop so slowly as to allow each intermediate step to be an equilibrium state are said to be reversible processes.

Thermodynamic Parameters

The central concept of thermodynamics is that of energy, the ability to do work. As stipulated by the first law, the total energy of the system and its surroundings is conserved. It may be transferred into a body by heating, compression, or addition of matter, and extracted from a body either by cooling, expansion, or extraction of matter.

For comparison, in mechanics, energy transfer results from a force which causes displacement, the product of the two being the amount of energy transferred. In a similar way, thermodynamic systems can be thought of as transferring energy as the result of a generalized force causing a generalized displacement, with the product of the two being the amount of energy transferred. These thermodynamic force-displacement pairs are known as conjugate variables. The most common conjugate thermodynamic variables are pressure-volume (mechanical parameters), temperature-entropy (thermal parameters), and chemical potential-particle number (material parameters).

Thermodynamic Instruments

There are two types of thermodynamic instruments, the meter and the reservoir. A thermodynamic meter is any device which measures any parameter of a thermodynamic system. In some cases, the thermodynamic parameter is actually defined in terms of an idealized measuring instrument. For example, the zeroth law states that if two bodies are in thermal equilibrium with a third body, they are also in thermal equilibrium with each other. This principle, as noted by James Maxwell in 1872, asserts that it is possible to measure temperature. An idealized thermometer is a sample of an ideal gas at constant pressure.

From the ideal gas law PV=nRT, the volume of such a sample can be used as an indicator of temperature; in this manner it defines temperature. Although pressure is defined mechanically, a pressure-measuring device, called a barometer may also be constructed from a sample of an ideal gas held at a constant temperature. A calorimeter is a device which is used to measure and define the internal energy of a system.

A thermodynamic reservoir is a system which is so large that it does not appreciably alter its state parameters when brought into contact with the test system. It is used to impose a particular value of a state parameter upon the system. For example, a pressure reservoir is a system at a particular pressure, which imposes that pressure upon any test system that it is mechanically connected to. The earth's atmosphere is often used as a pressure reservoir.

It is important that these two types of instruments are distinct. A meter does not perform its task accurately if it behaves like a reservoir of the state variable it is trying to measure. If, for example, a thermometer were to act as a temperature reservoir it would alter the temperature of the system being measured, and the reading would be incorrect. Ideal meters have no effect on the state variables of the system they are measuring.

Thermodynamic States

When a system is at equilibrium under a given set of conditions, it is said to be in a definite state. The state of the system can be described by a number of intensive variables and extensive variables. The properties of the system can be described by an equation of state which specifies the relationship between these variables. State may be thought of as the instantaneous quantitative description of a system with a set number of variables held constant

Thermodynamics is an experimental science based on a small number of principles that are generalizations made from experience. It is concerned only with macroscopic or large-scale properties of matter and it makes no hypotheses about the small-scale or microscopic structure of matter. From the principles of thermodynamics one can derive general relations between such quantities as coefficients of expansion, compressibilities, specific heat capacities, heats of transformation, and magnetic and dielectric coefficients, especially as these are affected by temperature. The principles of thermodynamics also tell us which of these relations must be determined experimentally in order to completely specify all the properties of the system...

Thermodynamics is complementary to kinetic theory and statistical thermodynamics. Thermodynamics provides relationships between physical properties of any system once certain measurements are made. Kinetic theory and statistical thermodynamics enable one to calculate the mangitudes of these properties for those systems whose energy states can be determined.

APPLICATION OF FIRST LAW OF HERMODYNAMICS

The Law of Conservation of Matter/Energy

The first law has been defined as follows: When a closed system is altered adiabatically, the total work associated with the change of state is the same for all possible processes between the two given equilibrium states.

A more succinct and comprehensible definition might be something like this: Matter/energy may be altered but not created (from nothingness) nor destroyed (reduced to nothingness). The First Law teaches that matter/energy cannot spring forth from nothing without cause, nor can it simply vanish.

The First Law, although not formally defined until the 19th century, helps make science possible. Science depends on the ability to identify cause-effect relationships. If matter/energy could spontaneously appear (and have effects on other matter/energy around it), scientists would never know whether a given observation was due to a rational cause, or to a spontaneous generation of matter or energy that was uncaused. Scientific conclusions would be on shaky ground. The Law of Causality is thus closely linked with the First Law of Thermodynamics.

The First Law also demands, if we accept it, one of two possibilities about the nature of the universe. One is that it has always existed, changing form perhaps but never having come from nothingness, or returning to the same. The other possibility is that it did not come from nothingness, but from a transcendant (that is, outside the universe) creator who is not subject to the laws within the universe.

First, they have unconsciously granted to the Law of Causality the very property of self-existence (that is, an eternal, uncreated nature) that they are presuming God couldn't have.

A being who created the universe and the laws within it, who pre-existed them, would not be slave to those laws. And since the Law of Causality is a statement about relationships between multiple entities, the law could not even exist until one entity began the act of creating another one (at which point it would implicitly come to exist).

Finally, most atheists who use this argument grant to the universe the exact property of self-existence that they deny God. They either deny the First Law of Thermodynamics and believe the universe came into existence from nothingness, or believe that it is itself self-existing. However, this latter position violates the unity principle – that a valid law of science that is found to apply anywhere, applies everywhere and to everything in the universe, including the universe as a whole.

The only position that appears to be consistent with the First Law of Thermodynamics, the unity principle and causality is that the universe was created by a self-existent external agent not subject to the laws operational in the universe it created.

THE SECOND LAW OF THERMODYNAMICS

In plain English the Second Law states that entropy always increases or remains constant in a closed system. (As a practical matter, for any non-trivial system entropy tends to increase due to irreversible processes.) The entropy of an entire closed system can never decrease within that system. Since the universe can be modeled as a closed system the universe is considered to be entropic – that is, running down.

The change in entropy (delta S) is equal to the heat transfer (delta Q) divided by the temperature (T).

delta S = (delta Q) / T

Second Law of Thermodynamics(heat engine): It is impossible to extract an amount of heat Q H from a hot reservoir and use it all to do work W . Some amount of heat Q C must be exhausted to a cold reservoir. This precludes a perfect heat engine.

Second Law of Thermodynamics(refrigerator): It is not possible for heat to flow from a colder body to a warmer body without any work having been done to accomplish this flow. Energy will not flow spontaneously from a low temperature object to a higher temperature object. This precludes a perfect refrigerator

THE THIRD LAW OF THERMODYNAMICS

The Third Law of Thermodynamics is the lesser known of the three major thermodynamic laws. Together, these laws help form the foundations of modern science. The laws of thermodynamics are absolute physical laws – everything in the observable universe is subject to them. Like time or gravity, nothing in the universe is exempt from these laws. In its simplest form, the Third Law of Thermodynamics relates the entropy (randomness) of matter to its absolute temperature.

The Third Law of Thermodynamics refers to a state known as "absolute zero." This is the bottom point on the Kelvin temperature scale. The Kelvin scale is absolute, meaning 0° Kelvin is mathematically the lowest possible temperature in the universe. This corresponds to about -273.15° Celsius, or -459.7 Fahrenheit .

In actuality, no object or system can have a temperature of zero Kelvin, because of the Second Law of Thermodynamics. The Second Law, in part, implies that heat can never spontaneously move from a colder body to a hotter body. So, as a system approaches absolute zero, it will eventually have to draw energy from whatever systems are nearby. If it draws energy, it can never obtain absolute zero. So, this state is not physically possible, but is a mathematical limit of the universe.

In its shortest form, the Third Law of Thermodynamics says: "The entropy of a pure perfect crystal is zero (0) at zero Kelvin (0° K)." Entropy is a property of matter and energy discussed by the Second Law of Thermodynamics. The Third Law of Thermodynamics means that as the temperature of a system approaches absolute zero, its entropy approaches a constant (for pure perfect crystals, this constant is zero). A pure perfect crystal is one in which every molecule is identical, and the molecular alignment is perfectly even throughout the substance. For non-pure crystals, or those with less-than perfect alignment, there will be some energy associated with the imperfections, so the entropy cannot become zero.

The Third Law of Thermodynamics can be visualized by thinking about water. Water in gas form has molecules that can move around very freely. Water vapor has very high entropy (randomness). As the gas cools, it becomes liquid. The liquid water molecules can still move around, but not as freely. They have lost some entropy.

When the water cools further, it becomes solid ice. The solid water molecules can no longer move freely, but can only vibrate within the ice crystals. The entropy is now very low. As the water is cooled more, closer and closer to absolute zero, the vibration of the molecules diminishes. If the solid water reached absolute zero, all molecular motion would stop completely. At this point, the water would have no entropy (randomness) at all.

TEMPERATURE

Temperature is the property of a body or region of space that determines whether or not there will be a net flow of heat into it or out of it from a neighboring body or region and in which direction the heat will flow. If there is no heat flow the bodies or regions are said to be in thermal equilibrium and at the same temperature.

If there is a flow of heat, the direction of the flow is from the body or region of higher temperature. Broadly, there are two methods of quantifying this property. The empirical method is to take two or more reproducible temperature-dependent events and assign fixed points on a scale of values to these events. For example, the Celsius temperature scale uses the freezing point and boiling point of water as the two fixed points, assigns the values 0 and 100 to them, respectively, and divides the scale between them into 100 degrees.

This method is serviceable for many practical purposes but lacking a theoretical basis it is awkward to use in many scientific contexts. In the 19th century, Lord Kelvin proposed a thermodynamic method to specify temperature, based on the measurement of the quantity of heat flowing between bodies at different temperatures. This concept relies on an absolute scale of temperature with an absolute zero of temperature, at which no body can give up heat. He also used Sadi Carnot's concept of an ideal frictionless perfectly efficient heat engine. This Carnot engine takes in a quantity of heat q1 at a temperature T1, and exhausts heat q2 at T2, so that T1/T2 = q1/q2. If T2 has a value fixed by definition, a Carnot engine can be run between this fixed temperature and any unknown temperature T1, enabling T1 to be calculated by measuring the values of q1 and q2. This concept remains the basis for defining thermodynamic temperature, quite independently of the nature of the working substance.

The unit in which thermodynamic temperature is expressed is the kelvin. In practice, thermodynamic temperatures cannot be measured directly; they are usually inferred from measurements with a gas thermometer containing a nearly ideal gas.

This is possible because another aspect of thermodynamic temperature is its relationship to the internal energy of a given amount of substance. This can be shown most simply in the case of an ideal monatomic gas, in which the internal energy per mole (U) is equal to the total kinetic energy of translation of the atoms in one mole of the gas (a monatomic gas has no rotational or vibrational energy). According to the kinetic theory, the thermodynamic temperature of such a gas is given by $T = 2U/3R$, where R is the universal gas constant.

Absolute Temperature

The Kelvin temperature scale (K) was developed by Lord Kelvin in the mid 1800s. The zero point of this scale is equivalent to -273.16 °C on the Celsius scale. This zero point is considered the lowest possible temperature of anything in the universe. Therefore, the Kelvin scale is also known as the "absolute temperature scale". At the freezing point of water, the temperature of the Kelvin scale reads 273 K. At the boiling point of water, it reads 373 K.

Whereas the Kelvin scale is widely used by scientists, the Celsius or Fahrenheit scales are used in daily life. These two scales are easier to understand than the large numbers of the Kelvin scale. Could you imagine waking up to your radio and hearing the DJ give a weather report like this: "It's going to be a beautiful day today with sunny skies and a balmy temperature of 297 K!" That's 24 °C or 75 °F .

Pressure-Volume (Constant Temperature)

What happens to the volume of a gas as the pressure on it changes. Let's try the following experiment using equipment that might be found in your kitchen.

Marshmallows are a mixture of sugar, air, and gelatin. The sugar makes them sweet, the air makes them fluffy, and the gelatin holds everything together. Marshmallows are a frozen foam and are mostly air by volume. When placed in a vacuum pump, they expand as the pressure decreases. Break the seal on their container and they shrink during the return to normal atmospheric pressure. Since the vacuum pump pulls on the marshmallows hard enough to burst some of the air bubbles, they are actually a bit smaller and more shriveled at the end of this experiment. The pressure of a gas is inversely proportional to its volume when temperature is constant. Symbolically .

$$P = 1/ V$$

This correlation was discovered independently by Robert Boyle of Ireland in 1662 and Edme Mariotte of France in 1676. In Great Britain , America , Australia , the West Indies and other remnants of the British Empire it is called Boyle's law , while in Continental Europe and other places it is called Mariotte's law .

Mariotte added the important provision that temperature remain constant. Boyle neglected to mention it, but the data he used to derive his law were most likely collected during a period in which the temperature did not experience any significant change.

Since the gas needs to be in thermal equilibrium with its environment (or some other heat reservoir) to maintain an even temperature, the pressure-volume relationship normally applies only to "slow" processes. The marshmallow-vacuum experiment shown above is an example of a "slow" process. The pressure is reduced at a rate slow enough that heat from the environment is able to keep the jar and its contents at nearly room temperature. Such a transformation that takes place without a change in temperature is said to be isothermal .

Pumping a bicycle tire with a hand pump is an example of a "fast" process. The work done pushing the piston transforms into an increase in the internal energy (and thus an increase in the temperature) of the air molecules within the pump. People familiar with hand bicycle pumps will attest to the fact that they get hot after use. Likewise, when a gas is allowed to expanded into a region of reduced pressure it does work on its surroundings.

The energy to do this work comes from the internal energy of the gas and so the temperature of the gas drops. You can experience this yourself without the aid of any apparatus other than your mouth. Purse your lips so that your mouth has only a tiny opening to the outside and blow hard. The air rushing from your mouth will be quite cool despite coming from the core of your body, which is normally quite hot (around 37 °C). During a "fast" process like the ones just described, pressure and volume are changing so rapidly that heat doesn't have enough time to get into or out of the gas to keep the temperature constant. Such a transformation that takes place without any flow of heat is said to be adiabatic

$$P1\ V1 = P2\ V2 = \text{constant}$$

Pressure Temperature (Constant Volume)

The pressure of a gas is directly proportional to its temperature when volume is constant. Symbolically .

$$P \propto T$$

An isochoric process is one that takes place without any change in volume .

This relationship doesn't really have a name, but I have heard it called the "pressure law" or (mistakenly) "Gay-Lussac's law".

$$\frac{p_1}{T_1} = \frac{p_2}{T_2} = \text{constant}$$

IDEAL GASES

An ideal gas is a substance possessing very simple thermodynamic properties to which actual gases and vapours appear to approximate indefinitely at low pressures and high temperatures. It has the characteristic equation pv=Re, and obeys Boyle's law at all temperatures. The coefficient of expansion at constant pressure is equal to the coefficient of increase of pressure at constant volume. The difference of the specific heats by equation is constant and equal to R. The isothermal elasticity – v(dp/dv) is equal to the pressure p. The adiabatic elasticity is equal to y p, where-y is the ratio S/s of the specific heats. The heat absorbed in isothermal expansion from vo to v at a temperature 0 is equal to the work done by equation (since d0 = o, and 0(dp/d0)dv =pdv), and both are given by the expression RO log e (v/vo).

The energy E and the total heat F are functions of the temperature only and their variations take the form dE = sdO, d F = Sd0. The specific heats are independent of the pressure or density by equations. If we also assume that they are constant with respect to temperature (which does not necessarily follow from the characteristic equation, but is generally assumed, and appears from Regnault's experiments to be approximately the case for simple gases), the expressions for the change of energy or total heat from 00 to 0 may be written E – E0 = s(0 – 0 0), F – Fo = S(0-00). In thiscase the ratio of the specific heats is constant as well as the difference, and the adiabatic equation takes the simple form, pv v = constant, which is at once obtained by integrating the equation for the adiabatic elasticity, – v(dp/dv) =yp.

DEVIATIONS OF ACTUAL GASES FROM THE IDEAL STATE

Since no gas is ideally perfect, it is most important for practical purposes to discuss the deviations of actual gases from the ideal state, and to consider how their properties may be thermodynamically explained and defined. The most natural method of procedure is to observe the deviations from Boyle's law by measuring the changes of pv at various constant temperatures. It is found by experiment that the change of pv with pressure at moderate pressures is nearly proportional to the change of p, in other words that the coefficient d(pv)/dp is to a first approximation a function of the temperature only.

This coefficient is sometimes called the " angular coefficient," and may be regarded as a measure of the deviations from Boyle's law, 'which may be most simply expressed at moderate pressures by formulating the variation of the angular coefficient with temperature. But this procedure in itself is not sufficient, because, although it would be highly probable that a gas obeying Boyle's law at all temperatures was practically an ideal gas, it is evident that Boyle's law would be satisfied by any substance having the characteristic equation pv = f (0), where f (0) is any arbitrary function of 0, and that the scale of temperatures given by such a substance would not necessarily coincide with the absolute scale.

A sufficient test, in addition to Boyle's law, is the condition dE/dv=o at constant temperature. This gives by equation the condition Odp/d0 =p, which is satisfied by any substance possessing the characteristic equation p/0=f(v), where f(v) is any arbitrary function of v. This test was applied by Joule in the well-known experiment in which he allowed a gas to expand from one vessel to another in a calorimeter without doing external work.

Under this condition the increase of intrinsic energy would be equal to the heat absorbed, and would be indicated by fall of temperature of the calorimeter. Joule failed to observe any change of temperature in his apparatus, and was therefore justified in assuming that the increase of intrinsic energy of a gas in isothermal expansion was very small, and that the absorption of heat observed in a similar experiment in which the gas was allowed to do external work by expanding against the atmospheric pressure was equivalent to the external work done.

But owing to the large thermal capacity of his calorimeter, the test, though sufficient for his immediate purpose, was not delicate enough to detect and measure the small deviations which actually exist.

JOULE–THOMSON EFFECT

In physics, the Joule–Thomson effect or Joule–Kelvin effect describes the temperature change of a gas or liquid when it is forced through a valve or porous plug while kept insulated so that no heat is exchanged with the environment.

This procedure is called a throttling process or Joule-Thomson process. At room temperature, all gases except hydrogen, helium and neon cool upon expansion by the Joule-Thomson process.

The effect is named for James Prescott Joule and William Thomson, 1st Baron Kelvin who discovered it in 1852 following earlier work by Joule on Joule expansion, in which a gas undergoes free expansion in a vacuum.

Description

The adiabatic (no heat exchanged) expansion of a gas may be carried out in a number of ways. The change in temperature experienced by the gas during expansion depends on the initial and final pressure, but also on the manner in which the expansion is carried out.

- If the expansion process is reversible, meaning that the gas is in thermodynamic equilibrium at all times, it is called an isentropic expansion. In this scenario, the gas does positive work during the expansion, and its temperature decreases.
- In a free expansion, on the other hand, the gas does no work and absorbs no heat, so the internal energy is conserved. Expanded in this manner, the temperature of an ideal gas would remain constant, but the temperature of a real gas may either increase or decrease, depending on the initial temperature and pressure.
- The method of expansion in which a gas or liquid at pressure P1 flows into a region of lower pressure P2 via a valve or porous plug under steady state conditions and without change in kinetic energy, is called the Joule–Thomson process. During this process, enthalpy remains unchanged.

Temperature change of either sign can occur during the Joule-Thomson process. Each real gas has a Joule–Thomson (Kelvin) inversion temperature above which expansion at constant enthalpy causes the temperature to rise, and below which such expansion causes cooling. This inversion temperature depends on pressure; for most gases at atmospheric pressure, the inversion temperature is above room temperature, so most gases can be cooled from room temperature by isenthalpic expansion.

Physical Mechanism

As a gas expands, the average distance between molecules grows. Because of intermolecular attractive forces expansion causes an increase in the potential energy of the gas. If no external work is extracted in the process and no heat is transferred, the total energy of the gas remains the same because of the conservation of energy. The increase in potential energy thus implies a decrease in kinetic energy and therefore in temperature.

A second mechanism has the opposite effect. During gas molecule collisions, kinetic energy is temporarily converted into potential energy. As the average intermolecular distance increases, there is a drop in the number of collisions per time unit, which causes a decrease in average potential energy. Again, total energy is conserved, so this leads to an increase in kinetic energy (temperature).

Below the Joule–Thomson inversion temperature, the former effect (work done internally against intermolecular attractive forces) dominates, and free expansion causes a decrease in temperature. Above the inversion temperature, gas molecules move faster and so collide more often, and the latter effect (reduced collisions causing a decrease in the average potential energy) dominates: Joule-Thomson expansion causes a temperature increase.

The Joule–Thomson (Kelvin) Coefficient

The rate of change of temperature T with respect to pressure P in a Joule–Thomson process (that is, at constant enthalpy H) is the Joule–Thomson (Kelvin) coefficient ìJT. This coefficient can be expressed in terms of the gas's volume V, its heat capacity at constant pressure Cp, and its coefficient of thermal expansion á as:

$$\mu_{JT} \equiv \left(\frac{\partial T}{\partial P}\right)_H = \frac{V}{C_p}(\alpha T - 1).$$

The value of μJT is typically expressed in °C/bar (SI units: K/Pa) and depends on the type of gas and on the temperature and pressure of the gas before expansion.

All real gases have an inversion point at which the value of μJT changes sign. The temperature of this point, the Joule–Thomson inversion temperature, depends on the pressure of the gas before expansion.

In a gas expansion the pressure decreases, so the sign of ∂P is always negative. With that in mind, the following table explains when the Joule–Thomson effect cools or warms a real gas:

If the gas temperature is	then μJT is	since ∂P is	thus∂P must be	so the gas cools
below the inversion temperature	positive	always negative	negative	
above the inversion temperature	negative	always negative	positive	warms

Helium and hydrogen are two gases whose Joule–Thomson inversion temperatures at a pressure of one atmosphere are very low (e.g., about 51 K ("222 °C) for helium). Thus, helium and hydrogen warm up when expanded at constant enthalpy at typical room temperatures. On the other hand nitrogen and oxygen, the two most abundant gases in air, have inversion temperatures of 621 K (348 °C) and 764 K (491 °C) respectively: these gases can be cooled from room temperature by the Joule–Thomson effect.

For an ideal gas, ìJT is always equal to zero: ideal gases neither warm nor cool upon being expanded at constant enthalpy.In practice, the Joule–Thomson effect is achieved by allowing the gas to expand through a throttling device (usually a valve) which must be very well insulated to prevent any heat transfer to or from the gas. No external work is extracted from the gas during the expansion. The effect is applied in the Linde technique as a standard process in the petrochemical industry, where the cooling effect is used to liquefy gases, and also in many cryogenic applications (e.g. for the production of liquid oxygen, nitrogen, and argon).

Only when the Joule–Thomson coefficient for the given gas at the given temperature is greater than zero can the gas be liquefied at that temperature by the Linde cycle. In other words, a gas must be below its inversion temperature to be liquefied by the Linde cycle. For this reason, simple Linde cycle liquefiers cannot normally be used to liquefy helium, hydrogen, or neon.

Proof that Enthalpy Remains Constant in a Joule-Thomson Process

In a Joule-Thomson process the enthalpy remains constant. To prove this, the first step is to compute the net work done by the gas that moves through the plug. Suppose that the gas has a volume of V1 in the region at pressure P1 and a volume of V2 when it appears in the region at pressure P2. Then the work done on the gas by the rest of the gas in region 1 is P1 V1. In region 2 the amount of work done by the gas is P2 V2. So, the total work done by the gas is

$$P_2V_2 - P_1V_1$$

The change in internal energy plus the work done by the gas is, by the first law of thermodynamics, the total amount of heat absorbed by the gas (here it is assumed that there is no change in kinetic energy).

In the Joule-Thompson process the gas is kept insulated, so no heat is absorbed. This means that

$$E_2 - E_1 + P_2V_2 - P_1V_1 = 0$$

where E1 and E2 denote the internal energy of the gas in regions 1 and 2, respectively.

The above equation then implies that:

$$H_1 = H_2$$

where H1 and H2 denote the enthalpy of the gas in regions 1 and 2, respectively.

Derivation of the Joule-Thomson (Kelvin) coefficient

A derivation of the formula

$$m_{JT} \equiv \left(\frac{\partial T}{\partial P}\right)_H = \frac{V}{C_P}(\alpha T - 1)$$

for the Joule–Thomson (Kelvin) coefficient.

The partial derivative of T with respect to P at constant H can be computed by expressing the differential of the enthalpy dH in terms of dT and dP, and equating the resulting expression to zero and solving for the ratio of dT and dP.

It follows from the fundamental thermodynamic relation that the differential of the enthalpy is given by:

$dH = TdS + VdP$ (here, S is the entropy of the gas).

Expressing dS in terms of dT and dP gives:

$$dH = T\left(\frac{\partial S}{\partial T}\right)_P dT + \left[V + T\left(\frac{\partial S}{\partial P}\right)_T\right]dP$$

Using

$$C_P = T\left(\frac{\partial S}{\partial T}\right)_P,$$ we can write:

$$dH = C_P dT + \left[V + T\left(\frac{\partial S}{\partial P}\right)_T\right]dP.$$

The remaining partial derivative of S can be expressed in terms of the coefficient of thermal expansion via a Maxwell relation as follows. From the fundamental thermodynamic relation, it follows that the differential of the Gibbs energy is given by:

$$dG = -SdT + VdP.$$

The symmetry of partial derivatives of G with respect to T and P implies that:

$$\left(\frac{\partial S}{\partial P}\right)_T = -\left(\frac{\partial V}{\partial T}\right)_P = -V\alpha$$

where α is the coefficient of thermal expansion. Using this relation, the differential of H can be expressed as

$$dH = C_p dT + V(1 - T\alpha)dP.$$

Equating dH to zero and solving for dT/dP then gives:

$$\left(\frac{\partial T}{\partial P}\right)_H = \frac{V}{C_P}(\alpha T - 1).$$

ENTROPY

It follows from the definition of the absolute scale of temperature, as given in relations that in passing at constant temperature 0 from one adiabatic 4' to any other adiabatic 0", the quotient H/o of the heat absorbed by the temperature at which it is absorbed is the same for the same two adiabatics whatever the temperature of the isothermal path. This quotient is called the change of entropy. In passing along an adiabatic there is no change of entropy, since no heat is absorbed. The adiabatics are lines of constant entropy, and are also called Isentropics.

In virtue of relations the change of entropy of a substance between any two states depends only on the initial and final states, and may be reckoned along any reversible path, not necessarily isothermal, by dividing each small increment of heat, dH, by the temperature, 0, at which it is acquired, and taking the sum or integral of the quotients, dH/o, so obtained.

In thermodynamics entropy, symbolized by S, is a measure of the unavailability of a system's energy to do work.It is a measure of the randomness of molecules in a system and is central to the second law of thermodynamics and the fundamental thermodynamic relation, which deal with physical processes and whether they occur spontaneously. Spontaneous changes, in isolated systems, occur with an increase in entropy.

Spontaneous changes tend to smooth out differences in temperature, pressure, density, and chemical potential that may exist in a system, and entropy is thus a measure of how far this smoothing-out process has progressed.

When a system's energy is defined as the sum of its "useful" energy, and its "useless energy", i.e. that energy which cannot be used for external work, then entropy may be visualized as the "scrap" or "useless" energy whose energetic prevalence over the total energy of a system is directly proportional to the absolute temperature of the considered system..

Entropy is a function of a quantity of heat which shows the possibility of conversion of that heat into work. The increase in entropy is small when heat is added at high temperature and is greater when heat is added at lower temperature. Thus for maximum entropy there is minimum availability for conversion into work and for minimum entropy there is maximum availability for conversion into work.

Entropy S is not defined directly, but rather by an equation relating the change in entropy of the system to the change in heat of the system.

For a constant temperature, the change in entropy ΔS is defined by the equation $\Delta S = \Delta Q/T$, where ΔQ is the amount of heat absorbed in an isothermal and reversible process in which the system goes from one state to another, and T is the absolute temperature at which the process is occurring. If the temperature of the system is not constant, then the relationship becomes a differential equation $dS = dQ/T$. To understand what this equation means, suppose the temperature T can be expressed as a function T(Q) of the heat Q . Then the total change in entropy as the heat-level varies is:

$$\Delta S = \int_A \frac{1}{T(Q)} dQ$$

where A is the set defining the range of heat values in the system.

Entropy is one of the factors that determines the free energy of the system. This thermodynamic definition of entropy is only valid for a system in equilibrium (because temperature is defined only for a system in equilibrium), while the statistical definition of entropy applies to any system. Thus the statistical definition is usually considered the fundamental definition of entropy.

Entropy increase has often been defined as a change to a more disordered state at a molecular level. In recent years, entropy has been interpreted in terms of the "dispersal" of energy. Entropy is an extensive state function that accounts for the effects of irreversibility in thermodynamic systems.In terms of statistical mechanics, the entropy describes the number of the possible microscopic configurations of the system. The statistical definition of entropy is the more fundamental definition, from which all other definitions and all properties of entropy follow.

In a thermodynamic system, a "universe" consisting of "surroundings" and "systems" and made up of quantities of matter, its pressure differences, density differences, and temperature differences all tend to equalize over time – simply because equilibrium state has higher probability than any other. In the ice melting example, the difference in temperature between a warm room (the surroundings) and cold glass of ice and water (the system and not part of the room), begins to be equalized as portions of the heat energy from the warm surroundings spread out to the cooler system of ice and water.

Over time the temperature of the glass and its contents and the temperature of the room become equal. The entropy of the room has decreased as some of its energy has been dispersed to the ice and water. However, as calculated in the example, the entropy of the system of ice and water has increased more than the entropy of the surrounding room has decreased. In an isolated system such as the room and ice water taken together, the dispersal of energy from warmer to cooler always results in a net increase in entropy.

Thus, when the 'universe' of the room and ice water system has reached a temperature equilibrium, the entropy change from the initial state is at a maximum. The entropy of the thermodynamic system is a measure of how far the equalization has progressed.

A special case of entropy increase, the entropy of mixing, occurs when two or more different substances are mixed. If the substances are at the same temperature and pressure, there will be no net exchange of heat or work – the entropy increase will be entirely due to the mixing of the different substances.

From a macroscopic perspective, in classical

thermodynamics the entropy is interpreted simply as a state function of a thermodynamic system: that is, a property depending only on the current state of the system, independent of how that state came to be achieved.

The state function has the important property that, when multiplied by a reference temperature, it can be understood as a measure of the amount of energy in a physical system that cannot be used to do thermodynamic work; i.e., work mediated by thermal energy.

More precisely, in any process where the system gives up energy ΔE, and its entropy falls by ΔS, a quantity at least TR ΔS of that energy must be given up to the system's surroundings as unusable heat (TR is the temperature of the system's external surroundings). Otherwise the process will not go forward.

Quantitatively, Clausius states the mathematical expression for this theorem is as follows. Let δQ be an element of the heat given up by the body to any reservoir of heat during its own changes, heat which it may absorb from a reservoir being here reckoned as negative, and T the absolute temperature of the body at the moment of giving up this heat, then the equation:

$$\int \frac{\delta Q}{T} = 0$$

must be true for every reversible cyclical process, and the relation:

$$\int \frac{\delta Q}{T} \geq 0$$

must hold good for every cyclical process which is in any way possible. This is the essential formulation of the second law and one of the original forms of the concept of entropy. The dimensions of entropy are energy divided by temperature, which is the same as the dimensions of Boltzmann's constant (kB) and heat capacity. The SI unit of entropy is "joule per kelvin" (J·K"1). In this manner, the quantity "ΔS" is utilized as a type of internal energy, which accounts for the effects of irreversibility, in the energy balance equation for any given system

In the Gibbs free energy equation, i.e. ΔG = ΔH – TΔS, for example, which is a formula commonly utilized to determine if chemical reactions will occur, the energy related to entropy changes TΔS is subtracted from the "total" system energy ΔH to give the "free" energy ΔG of the system, as during a chemical process or as when a system changes state.

Microscopic Definition of Entropy (Statistical Mechanics)

In statistical thermodynamics the entropy is defined as (proportional to) the logarithm of the number of microscopic configurations that result in the observed macroscopic description of the thermodynamic system:

$$S = k_B \ln \Omega$$

where

kB is Boltzmann's constant 1.38066×10–23 J K–1 and

Ωis the number of microstates corresponding to the observed thermodynamic macrostate.

This definition is considered to be the fundamental definition of entropy. In Boltzmann's 1896 Lectures on Gas Theory, he showed that this expression gives a measure of entropy for systems of atoms and molecules in the gas phase, thus providing a measure for the entropy of classical thermodynamics.

In 1877, Boltzmann visualized a probabilistic way to measure the entropy of an ensemble of ideal gas particles, in which he defined entropy to be proportional to the logarithm of the number of microstates such a gas could occupy. Henceforth, the essential problem in statistical thermodynamics, i.e. according to Erwin Schrödinger, has been to determine the distribution of a given amount of energy E over N identical systems.

Statistical mechanics explains entropy as the amount of uncertainty which remains about a system, after its observable macroscopic properties have been taken into account. For a given set of macroscopic variables, like temperature and volume, the entropy measures the degree to which the probability of the system is spread out over different possible quantum states. The more states available to the system with higher probability, the greater the entropy. More specifically, entropy is a logarithmic measure of the density of states. In essence, the most general interpretation of entropy is as a measure of our uncertainty about a system. The equilibrium state of a system maximizes the entropy because we have lost all information about the initial conditions except for the conserved variables; maximizing the entropy maximizes our ignorance about the details of the system. This uncertainty is not of the everyday subjective kind, but rather the uncertainty inherent to the experimental method and interpretative model.

On the molecular scale, the two definitions match up because adding heat to a system, which increases its classical thermodynamic entropy, also increases the system's thermal fluctuations, so giving an increased lack of information about the exact microscopic state of the system, i.e. an increased statistical mechanical entropy.

The interpretative model has a central role in determining entropy. The qualifier "for a given set of macroscopic variables" above has very deep implications: if two observers use different sets of macroscopic variables, then they will observe different entropies. For example, if observer A uses the variables U, V and W, and observer B uses U, V, W, X, then, by changing X, observer B can cause an effect that looks like a violation of the second law of thermodynamics to observer A. In other words: the set of macroscopic variables one chooses must include everything that may change in the experiment, otherwise one might see decreasing entropy.

Entropy in Chemical Thermodynamics

Thermodynamic entropy is central in chemical thermodynamics, enabling changes to be quantified and the outcome of reactions predicted. The second law of thermodynamics states that entropy in the combination of a system and its surroundings increases during all spontaneous chemical and physical processes. Spontaneity in chemistry means "by itself, or without any outside influence", and has nothing to do with speed. The Clausius equation of $\delta q_{rev}/T = \Delta S$ introduces the measurement of entropy change, ΔS. Entropy change describes the direction and quantitates the magnitude of simple changes such as heat transfer between systems – always from hotter to cooler spontaneously. Thus, when a mole of substance at 0 K is warmed by its surroundings to 298 K, the sum of the incremental values of qrev/T constitute each element's or compound's standard molar entropy, a fundamental physical property and an indicator of the amount of energy stored by a substance at 298 K. Entropy change also measures the mixing of substances as a summation of their relative quantities in the final mixture.

Entropy is equally essential in predicting the extent of complex chemical reactions, i.e. whether a process will go as written or proceed in the opposite direction. For such applications, ΔS must be incorporated in an expression that includes both the system and its surroundings, ΔSuniverse = ΔSsurroundings + ΔS system. This expression becomes, via some steps, the Gibbs free energy equation for reactants and products in the system: ΔG [the Gibbs free energy change of the system] = ΔH [the enthalpy change] "T ΔS [the entropy change].

Entropy Balance Equation for Open Systems

In chemical engineering, the principles of thermodynamics are commonly applied to "open systems", i.e. those in which heat, work, and mass flow across the system boundary. In a system in which there are flows of both heat ($\dot{Q}$) and work, i.e. $\dot{W}_S$(shaft work) and P(dV/dt) (pressure-volume work), across the system boundaries, the heat flow, but not the work flow, causes a change in the entropy of the system.

This rate of entropy change is $\dot{Q}/T$ where T is the absolute thermodynamic temperature of the system at the point of the heat flow.

If, in addition, there are mass flows across the system boundaries, the total entropy of the system will also change due to this convected flow.

To derive a generalized entropy balanced equation, we start with the general balance equation for the change in any extensive quantity Θ in a thermodynamic system, a quantity that may be either conserved, such as energy, or non-conserved, such as entropy.

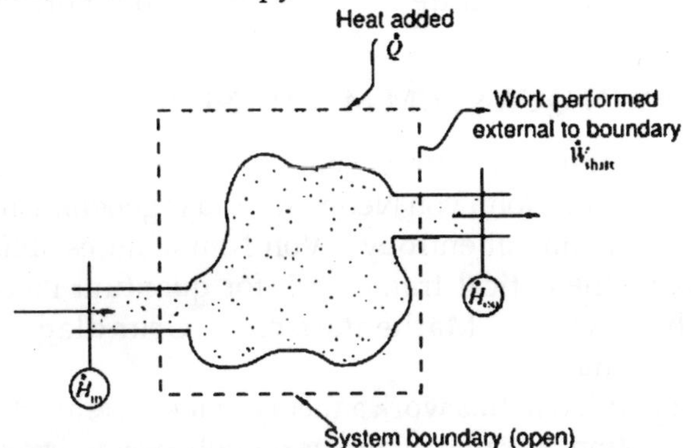

Fig. Entropy Balance Applied to An Open System

The basic generic balance expression states that dΘ/dt, i.e. the rate of change of Θ in the system, equals the rate at which Θ enters the system at the boundaries, minus the rate at which Θ leaves the system across the system boundaries, plus the rate at which Θ is generated within the system.

Using this generic balance equation, with respect to the rate of change with time of the extensive quantity entropy S, the entropy balance equation for an open thermodynamic system is:

$$\frac{dS}{dt} = \sum_{k=1}^{K} \dot{M}_k \hat{S}_k + \frac{\dot{Q}}{T} + \dot{S}_{gen}$$

where

$\sum_{k=1}^{K} \dot{M}_k \hat{S}_k$ = the net rate of entropy flow due to the flows of mass into and out of the system (where $\hat{S}$ = entropy per unit mass).

$\frac{\dot{Q}}{T}$ = the rate of entropy flow due to the flow of heat across the system boundary.

$\dot{S}_{gen}$ = the rate of internal generation of entropy within the system.

Note, also, that if there are multiple heat flows, the term $\dot{Q}/T$ is to be replaced by $\sum \dot{Q}_j / T_j$, where $\dot{Q}_j$ is the heat flow and Tj is the temperature at the jth heat flow port into the system.

ENTROPY IN QUANTUM MECHANICS

In quantum statistical mechanics, the concept of entropy was developed by John von Neumann and is generally referred to as "von Neumann entropy". Von Neumann established a rigorous mathematical framework for quantum mechanics with his work Mathematische Grundlagen der Quantenmechanik.

He provided in this work a theory of measurement, where the usual notion of wave collapse is described as an irreversible process (the so called von Neumann or projective measurement).

Using this concept, in conjunction with the density matrix he extended the classical concept of entropy into the quantum domain.

It is well known that a Shannon based definition of information entropy leads in the classical case to the Boltzmann entropy.

It is tempting to regard the Von Neumann entropy as the corresponding quantum mechanical definition. But the latter is problematic from quantum information point of view.

Consequently Stotland, Pomeransky, Bachmat and Cohen have introduced a new definition of entropy that reflects the inherent uncertainty of quantum mechanical states.

This definition allows to distinguish between the minimum uncertainty entropy of pure states, and the excess statistical entropy of mixtures.

Entropy in Astrophysics

In astrophysics, what is referred to as "entropy" is actually the adiabatic constant derived as follows.

Using the first law of thermodynamics for a quasi-static, infinitesimal process for a hydrostatic system

$$dQ = dU - dW.$$

For an ideal gas in this special case, the internal energy, U, is only a function of T; therefore the partial derivative of heat capacity with respect to T is identically the same as the full derivative, yielding through some manipulation

$$dQ = C_V dT + PdV.$$

Further manipulation using the differential version of the ideal gas law, the previous equation, and assuming constant pressure, one finds

$$dQ = C_P dT - VdP.$$

For an adiabatic process dQ = 0 and recalling $\gamma = \frac{C_P}{C_V}$, one finds

$$VdP = C_P dT$$
$$PdV = -C_V dT$$
$$\frac{dP}{P} = -\frac{dV}{P}\gamma.$$

One can solve this simple differential equation to find

$$PV^{\gamma} = \text{constant} = K$$

This equation is known as an expression for the adiabatic constant, K, also called the adiabat. From the ideal gas equation one also knows

$$P = \frac{\rho k_B T}{\mu m_H},$$

where k_B is Boltzmann's constant. Substituting this into the above equation along with $V = [grams]/\rho$ and $\gamma = 5/3$ for an ideal monoatomic gas one finds

$$K = \frac{k_B T}{\mu m_H r^{2/3}},$$

where μ is the mean molecular weight of the gas or plasma; and μ_H is the mass of the Hydrogen atom, which is extremely close to the mass of the proton, m_p, the quantity more often used in astrophysical theory of galaxy clusters. This is what astrophysicists refer to as "entropy" and has units of [keV cm2]. This quantity relates to the thermodynamic entropy as

$$S = k_B \ln \Omega + S_0$$

where Ω, the density of states in statistical theory, takes on the value of K as defined above.

Energy Dispersal

The concept of entropy can be described qualitatively as a measure of energy dispersal at a specific temperature. Similar terms have been in use from early in the history of classical thermodynamics, and with the development of statistical thermodynamics and quantum theory, entropy changes have been described in terms of the mixing or "spreading" of the total energy of each constituent of a system over its particular quantized energy levels.

Ambiguities in the terms disorder and chaos, which usually have meanings directly opposed to equilibrium, contribute to widespread confusion and hamper comprehension of entropy for most students. As the second law of thermodynamics shows, in an isolated system internal portions at different temperatures will tend to adjust to a single uniform temperature and thus produce equilibrium.

A recently developed educational approach avoids ambiguous terms and describes such spreading out of energy as dispersal, which leads to loss of the differentials required for work even though the total energy remains constant in accordance with the first law of thermodynamics. Physical chemist Peter Atkins, for example, who previously wrote of dispersal leading to a disordered state, now writes that "spontaneous changes are always accompanied by a dispersal of energy", and has discarded 'disorder' as a description.

Entropy and Information Theory

In information theory, entropy is the measure of the amount of information that is missing before reception and is sometimes referred to as Shannon entropy. Shannon entropy is a broad and general concept which finds applications in information theory as well as thermodynamics. It was originally devised by Claude Shannon in 1948 to study the amount of information in a transmitted message. The definition of the information entropy is, however, quite general, and is expressed in terms of a discrete set of probabilities pi. In the case of transmitted messages, these probabilities were the probabilities that a particular message was actually transmitted, and the entropy of the message system was a measure of how much information was in the message. For the case of equal probabilities (i.e. each message is equally probable), the Shannon entropy (in bits) is just the number of yes/no questions needed to determine the content of the message.

The question of the link between information entropy and thermodynamic entropy is a hotly debated topic. Some authors argue that there is a link between the two, while others will argue that they have absolutely nothing to do with each other.

The expressions for the two entropies are very similar. The information entropy H for equal probabilities pi = p is:

$$H = K \ln (1/p)$$

where K is a constant which determines the units of entropy. For example, if the units are bits, then K=1/ln. The thermodynamic entropy S , from a statistical mechanical point of view was first expressed by Boltzmann:

$$S = 1 \ln (1/p)$$

where p is the probability of a system being in a particular microstate, given that it is in a particular macrostate, and k is Boltzmann's constant. It can be seen that one may think of the thermodynamic entropy as Boltzmann's constant, divided by ln, times the number of yes/no questions that must be asked in order to determine the microstate of the system, given that we know the macrostate.

The link between thermodynamic and information entropy was developed in a series of papers by Edwin Jaynes beginning in 1957.

The problem with linking thermodynamic entropy to information entropy is that in information entropy the entire body of thermodynamics which deals with the physical nature of entropy is missing.

The second law of thermodynamics which governs the behavior of thermodynamic systems in equilibrium, and the first law which expresses heat energy as the product of temperature and entropy are physical concepts rather than informational concepts.

If thermodynamic entropy is seen as including all of the physical dynamics of entropy as well as the equilibrium statistical aspects, then information entropy gives only part of the description of thermodynamic entropy. Some authors, like Tom Schneider, argue for dropping the word entropy for the H function of information theory and using Shannon's other term "uncertainty" instead.

Ice Melting Example

The illustration for this article is a classic example in which entropy increases in a small 'universe', a thermodynamic system consisting of the 'surroundings' (the warm room) and 'system' (glass, ice, cold water). In this universe, some heat energy δQ from the warmer room surroundings (at 298 K or 25 °C) will spread out to the cooler system of ice and water at its constant temperature T of 273 K (0°C), the melting temperature of ice.

The entropy of the system will change by the amount dS = δQ/T, in this example δQ/273 K. (The heat δQ for this process is the energy required to change water from the solid state to the liquid state, and is called the enthalpy of fusion, i.e. the ΔH for ice fusion.) The entropy of the surroundings will change by an amount dS = "δQ/298 K. So in this example, the entropy of the system increases, whereas the entropy of the surroundings decreases.

It is important to realize that the decrease in the entropy of the surrounding room is less than the increase in the entropy of the ice and water: the room temperature of 298 K is larger than 273 K and therefore the ratio, (entropy change), of δQ/298 K for the surroundings is smaller than the ratio (entropy change), of δQ/273 K for the ice+water system.

To find the entropy change of our "universe", we add up the entropy changes for its constituents: the surrounding room and the ice+water. The total entropy change is positive; this is always true in spontaneous events in a thermodynamic system and it shows the predictive importance of entropy: the final net entropy after such an event is always greater than was the initial entropy.

As the temperature of the cool water rises to that of the room and the room further cools imperceptibly, the sum of the $\delta Q/T$ over the continuous range, at many increments, in the initially cool to finally warm water can be found by calculus. The entire miniature "universe", i.e. this thermodynamic system, has increased in entropy. Energy has spontaneously become more dispersed and spread out in that "universe" than when the glass of ice water was introduced and became a "system" within it.

Entropy and Cosmology

As a finite universe may be considered an isolated system, it may be subject to the Second Law of Thermodynamics, so that its total entropy is constantly increasing. It has been speculated that the universe is fated to a heat death in which all the energy ends up as a homogeneous distribution of thermal energy, so that no more work can be extracted from any source.

If the universe can be considered to have generally increasing entropy, then – as Roger Penrose has pointed out – gravity plays an important role in the increase because gravity causes dispersed matter to accumulate into stars, which collapse eventually into black holes.

Jacob Bekenstein and Stephen Hawking have shown that black holes have the maximum possible entropy of any object of equal size. This makes them likely end points of all entropy-increasing processes, if they are totally effective matter and energy traps. Hawking has, however, recently changed his stance on this aspect.

The role of entropy in cosmology remains a controversial subject. Recent work has cast extensive doubt on the heat death hypothesis and the applicability of any simple thermodynamic model to the universe in general.

Although entropy does increase in the model of an expanding universe, the maximum possible entropy rises much more rapidly – thus entropy density is decreasing with time. This results in an "entropy gap" pushing the system further away from equilibrium. Other complicating factors, such as the energy density of the vacuum and macroscopic quantum effects, are difficult to reconcile with thermodynamical models, making any predictions of large-scale thermodynamics extremely difficult.

IRREVERSIBLE PROCESSES

In order that a process may be strictly reversible, it is necessary that the state of the working substance should be one of equilibrium at uniform pressure and temperature throughout.

If heat passes " of itself " from a higher to a lower temperature by conduction, convection or radiation, the transfer cannot be reversed without an expenditure of work. If mechanical work or kinetic energy is directly converted into heat by friction, reversal of the motion does not restore the energy so converted. In all such cases there is necessarily, by Carnot's principle, a loss of efficiency or available energy, accompanied by an increase of entropy, which serves as a convenient measure or criterion of the loss.

A common illustration of an irreversible process is the expansion of a gas into a vacuum or against a pressure less than its own. In this case the work of expansion, pdv, is expended in the first instance in producing kinetic energy of motion of parts of the gas. If this could be co-ordinated and utilized without dissipation, the gas might conceivably be restored to its initial state; but in practice violent local differences of pressure and temperature are produced, the kinetic energy is rapidly converted into heat by viscous eddy friction, and residual differences of temperature are equalized by diffusion throughout the mass.

Even if the expansion is adiabatic, in the sense that it takes place inside a non-conducting enclosure and no heat is supplied from external sources, it will not be isentropic, since the heat supplied by internal friction must be included in reckoning the change of entropy. Assuming that no heat is supplied from external sources and no external work is done, the intrinsic energy remains constant by the first law. The final state of the substance, when equilibrium has been restored, may be deduced from this condition, if the energy can be expressed in terms of the co-ordinates.

But the line of constant energy on – the diagram does not represent the path of the transformation, unless it be supposed to be effected in a series of infinitesimal steps between each of which the substance is restored to an equilibrium state. An irreversible process which permits a more complete experimental investigation is the steady flow of a fluid in a tube .

If the tube is a perfect non-conductor, and if there are no eddies or frictional dissipation, the state of the substance at any point of the tube as to E, p, and v, is represented by the adiabatic or isentropic path, dE= -pdv. The tube varies, the change of kinetic energy of flow, dU, is represented by The flow in this case is reversible, and the state of the fluid is the same at points where the section of the tube is the same.

In practice, however, there is always some frictional dissipation, accompanied by an increase of entropy and by a fall of pressure. In the limiting case of a long fine tube, the bore of which varies in such a manner that U is constant, the state of the substance along a line of flow may be represented by the line of constant total heat, d(E+pv) = o; but in the case of a porous plug or small throttling aperture, the steps of the process cannot be followed, though the final state is the same.

Chapter 2

Heat and Work

This theory, known to history as the phlogiston theory, is so extraordinary stupid that it is not even worth describing. In place of phlogiston theory, Lavoisier proposed the first reasonably sensible scientific interpretation of heat. Lavoisier pictured heat as an invisible, tasteless, odourless, weightless fluid, which he called calorific fluid. He postulated that hot bodies contain more of this fluid than cold bodies. Furthermore, he suggested that the constituent particles of calorific fluid repel one another, causing heat to flow from hot to cold bodies when they are placed in thermal contact.

The modern interpretation of heat is, or course, somewhat different to Lavoisier's calorific theory. Nevertheless, there is an important subset of problems involving heat flow for which Lavoisier's approach is rather useful. These problems often crop up as examination questions.

For example: "A clean dry copper calorimeter contains 100 grams of water at 30° degrees centigrade. A 10 gram block of copper heated to 60° centigrade is added. What is the final temperature of the mixture?". How do we approach this type of problem?

Well, according to Lavoisier's theory, there is an analogy between heat flow and incompressible fluid flow under gravity. The same volume of liquid added to containers of different cross-sectional area fills them to different heights. If the volume is V, and the cross-sectional area is A, then the height is $h = V/A$. In a similar manner, the same quantity of heat added to different bodies causes them to rise to different temperatures. If Q is the heat and θ is the (absolute) temperature then $\theta = Q/C$, where the constant C is termed the heat capacity. [This is a somewhat oversimplified example. In general, the heat capacity is a function of temperature, so that $C = C(\theta)$]. Now, if two containers filled to different heights with a free flowing incompressible fluid are connected together at the bottom, via a small pipe, then fluid will flow under gravity, from one to the other, until the two heights are the same. The final height is easily calculated by equating the total fluid volume in the initial and final states. Thus,

$$h_1A_1 + h_2A_2 = hA_1 + hA_2,$$

giving

$$h = \frac{h_1A_1 + h_2A_2}{A_1 + A_2}.$$

Here, h_1 and h_2 are the initial heights in the two containers, $A_1$and A_2 are the corresponding cross-sectional areas, and is the final height. Likewise, if two bodies, initially at different temperatures, are brought into thermal contact then heat will flow, from one to the other, until the two temperatures are the same. The final temperature is calculated by equating the total heat in the initial and final states. Thus,

$$\theta_1C_1 + \theta_2C_2 = \theta C_1 + \theta C_2,$$

giving

$$\theta = \frac{\theta_1C_1 + \theta_2C_2}{C_1 + C_2},$$

where the meaning of the various symbols should be self-evident.

The analogy between heat flow and fluid flow works because in Lavoisier's theory heat is a conserved quantity, just like the volume of an incompressible fluid. In fact, Lavoisier postulated that heat was an element. Note that atoms were thought to be indestructible before nuclear reactions were discovered, so the total amount of each element in the Universe was assumed to be a constant. If Lavoisier had cared to formulate a law of thermodynamics from his calorific theory then he would have said that the total amount of heat in the Universe was a constant.

In 1798 Benjamin Thompson, an Englishman who spent his early years in pre-revolutionary America, was minister for war and police in the German state of Bavaria. One of his jobs was to oversee the boring of cannons in the state arsenal. Thompson was struck by the enormous, and seemingly inexhaustible, amount of heat generated in this process. He simply could not understand where all this heat was coming from.

According to Lavoisier's calorific theory, the heat must flow into the cannon from its immediate surroundings, which should, therefore, become colder. The flow should also eventually cease when all of the available heat has been extracted. In fact, Thompson observed that the surroundings of the cannon got hotter, not colder, and that the heating process continued unabated as long as the boring machine was operating. Thompson postulated that some of the mechanical work done on the cannon by the boring machine was being converted into heat. At the time, this was quite a revolutionary concept, and most people were not ready to accept it.

This is somewhat surprising, since by the end of the eighteenth century the conversion of heat into work, by steam engines, was quite commonplace. Joule confirmed that work could indeed be converted into heat. Moreover, he found that the same amount of work always generates the same quantity of heat.

This is true regardless of the nature of the work (e.g., mechanical, electrical, etc.). Joule was able to formulate what became known as the work equivalent of heat. Namely, that 1 newton meter of work is equivalent to 0.241 calories of heat. A calorie is the amount of heat required to raise the temperature of 1 gram of water by 1 degree centigrade. Nowadays, we measure both heat and work in the same units, so that one newton meter, or joule, of work is equivalent to one joule of heat.

In 1850, the German physicist Clausius correctly postulated that the essential conserved quantity is neither heat nor work, but some combination of the two which quickly became known as energy, from the Greek energia meaning "in work." According to Clausius, the change in the internal energy of a macroscopic body can be written

$$\Delta E = Q - W,$$

where Q is the heat absorbed from the surroundings, and W is the work done on the surroundings. This relation is known as the first law of thermodynamics.

MACROSTATES AND MICROSTATES

In describing a system made up of a great many particles, it is usually possible to specify some macroscopically measurable independent parameters $x_1 x_2, \ldots x_n$, which affect the particles' equations of motion. These parameters are termed the external parameters the system. Examples of such parameters are the volume (this gets into the equations of motion because the potential energy becomes infinite when a particle strays outside the available volume) and any applied electric and magnetic fields. A microstate of the system is defined as a state for which the motions of the individual particles are completely specified (subject, of course, to the unavoidable limitations imposed by the uncertainty principle of quantum mechanics). In general, the overall energy of a given microstate is a function of the external parameters:

$$E_r \equiv E_r(x_1, x_2, \ldots, x_n).$$

A macrostate of the system is defined by specifying the external parameters, and any other constraints to which the system is subject. For example, if we are dealing with an isolated system (i.e., one that can neither exchange heat with nor do work on its surroundings) then the macrostate might be specified by giving the values of the volume and the constant total energy. For a many-particle system, there are generally a very great number of microstates which are consistent with a given macrostate.

THE MICROSCOPIC INTERPRETATION OF HEAT AND WORK

Consider a macroscopic system A which is known to be in a given macrostate. To be more exact, consider an ensemble of similar macroscopic systems A, where each system in the ensemble is in one of the many microstates consistent with the given macrostate. There are two fundamentally different ways in which the average energy of A can change due to interaction with its surroundings. If the external parameters of the system remain constant then the interaction is termed a purely thermal interaction. Any change in the average energy of the system is attributed to an exchange of heat with its environment. Thus,

$$\Delta\bar{E} = Q,$$

where Q is the heat absorbed by the system. On a microscopic level, the energies of the individual microstates are unaffected by the absorption of heat. In fact, it is the distribution of the systems in the ensemble over the various microstates which is modified.

Suppose that the system A is thermally insulated from its environment. This can be achieved by surrounding it by an adiabatic envelope (i.e., an envelope fabricated out of a material which is a poor conductor of heat, such a fiber glass). Incidentally, the term adiabatic is derived from the Greek adiabatos which means "impassable." In scientific terminology, an adiabatic process is one in which there is no exchange of heat. The system A is still capable of interacting with its environment via its external parameters. This type of interaction is termed mechanical interaction, and any change in the average energy of the system is attributed to work done on it by its surroundings. Thus,

$$\Delta\bar{E} = -W,$$

where is the work done by the system on its environment. On a microscopic level, the energy of the system changes because the energies of the individual microstates are functions of the external parameters. Thus, if the external parameters are changed then, in general, the energies of all of the systems in the ensemble are modified (since each is in a specific microstate). Such a modification usually gives rise to a redistribution of the systems in the ensemble over the accessible microstates (without any heat exchange with the environment). Clearly, from a microscopic viewpoint, performing work on a macroscopic system is quite a complicated process. Nevertheless, macroscopic work is a quantity which can be readily measured experimentally. For instance, if the system A exerts a force F on its immediate surroundings, and the change in external parameters corresponds to a displacement x of the center of mass of the system, then the work done by A on its surroundings is simply

$$W = F.x$$

i.e., the product of the force and the displacement along the line of action of the force. In a general interaction of the system A with its environment there is both heat exchange and work performed. We can write

$$Q \equiv \Delta\bar{E} + W, 7$$

which serves as the general definition of the absorbed heat Q(hence, the equivalence sign). The quantity Q is simply the change in the mean energy of the system which is not due to the modification of the external parameters. Note that the notion of a quantity of heat has no independent meaning apart from Eq. The mean energy $\bar{E}$ and work performed are both physical quantities which can be determined experimentally, whereas Q is merely a derived quantity.

QUASI-STATIC PROCESSES

Consider the special case of an interaction of the system A with its surroundings which is carried out so slowly that A remains arbitrarily close to equilibrium at all times. Such a process is said to be quasi-static for the system A. In practice, a quasi-static process must be carried out on a time-scale which is much longer than the relaxation time of the system. Recall that the relaxation time is the typical time-scale for the system to return to equilibrium after being suddenly disturbed.

A finite quasi-static change can be built up out of many infinitesimal changes. The infinitesimal heat $đW$ absorbed by the system when infinitesimal work $đW$ is done on its environment and its average energy changes by $d\bar{E}$ is given by

$$đQ \equiv dE + đW.$$

The special symbols $đW$ and $đQ$ are introduced to

emphasize that the work done and heat absorbed are infinitesimal quantities which do not correspond to the difference between two works or two heats. Instead, the work done and heat absorbed depend on the interaction process itself. Thus, it makes no sense to talk about the work in the system before and after the process, or the difference between these.

If the external parameters of the system have the values $x_1, \ldots, x_n$, then the energy of the system in a definite microstate can be written

$$E_r = E_r(x_1, \cdots, x_n).$$

Hence, if the external parameters are changed by infinitesimal amounts, so that $x_\alpha \to x_\alpha + dx_\alpha$ for in the range 1 to , then the corresponding change in the energy of the microstate is

$$dE_r = \sum_{a=1}^{n} \frac{\partial E_r}{\partial x_\alpha} dx_\alpha .$$

The work $\bar{d}W$ done by the system when it remains in this particular state r is

$$\bar{d}W_r = -dE_r = \sum_{\alpha=1}^{n} X_{\alpha r} dx_\alpha ,$$

where

$$X_{\alpha r} \equiv -\frac{\partial E_r}{\partial x_\alpha}$$

is termed the generalized force (conjugate to the external parameter x_α) in the state r. Note that if is x_α a displacement then $X_{\alpha r}$ is an ordinary force.

Consider now an ensemble of systems. Provided that the external parameters of the system are changed quasi-statically, the generalized forces $X_{\alpha r}$ have well defined mean values which are calculable from the distribution of systems in the ensemble characteristic of the instantaneous macrostate. The macroscopic work $\bar{d}W$ resulting from an infinitesimal quasi-static change of the external parameters is obtained by calculating the decrease in the mean energy resulting from the parameter change. Thus,

$$\bar{d}W = \sum_{\alpha=1}^{n} \bar{X}_{\alpha} dx_{\alpha},$$

where

$$\bar{X}_{\alpha} \equiv -\overline{\frac{\partial E_r}{\partial x_{\alpha}}}$$

is the mean generalized force conjugate to x_{α}. The mean value is calculated from the equilibrium distribution of systems in the ensemble corresponding to the external parameter values x_{α}. The macroscopic work W resulting from a finite quasi-static change of external parameters can be obtained by integrating Eq.

The most well-known example of quasi-static work in thermodynamics is that done by pressure when the volume changes. For simplicity, suppose that the volume V is the only external parameter of any consequence. The work done in changing the volume from V to $V + dV$ is simply the product of the force and the displacement (along the line of action of the force). By definition, the mean equilibrium pressure $\bar{p}$ of a given macrostate is equal to the normal force per unit area acting on any surface element. Thus, the normal force acting on a surface element dS_i is $\bar{p}\, dS_i$. Suppose that the surface element is subject to a displacement dx_i. The work done by the element is $\bar{p}\, dS_i.dx_i$. The total work done by the system is obtained by summing over all of the surface elements. Thus,

$$\bar{d}W = \bar{p}\, dV,$$

where

$$dV = \sum_i dS_i.dx_i$$

is the infinitesimal volume change due to the displacement of the surface. It follows from that

$$\bar{p} = -\frac{\partial \bar{E}}{\partial V},$$

so the mean pressure is the generalized force conjugate to the volume V. Suppose that a quasi-static process is carried out in which the volume is changed from V_i to V_f. In general, the mean pressure is a function of the volume, so $\bar{p} = \bar{p}(V)$. It follows that the macroscopic work done by the system is given by

$$W_{if} = \int_{V_i}^{V_f} đW = \int_{V_i}^{V_f} \bar{p}(V)\, dV.$$

This quantity is just the "area under the curve" in a plot of $\bar{p}(V)$ versus V.

EXACT AND INEXACT DIFFERENTIALS

In our investigation of heat and work we have come across various infinitesimal objects such as $d\bar{E}$ and $đW$. It is instructive to examine these infinitesimals more closely.

Consider the purely mathematical problem where $F(x, y)$is some general function of two independent variables x and y. Consider the change in F in going from the point (x, y)in the x–y plane to the neighbouring point $(x + dx, y + dy)$. This is given by

$$dF = F(x+dx, y+dy) - F(x,y),$$

which can also be written

$$dF = X(x, y)dx + Y(x, y)dy,$$

where $X = \partial F/\partial x$ and $Y=\partial F/\partial y$. Clearly, dF is simply the infinitesimal difference between two adjacent values of the function F. This type of infinitesimal quantity is termed an exact differential to distinguish it from another type to be discussed presently. If we move in the x–y plane from an initial point $i \equiv (x_i, y_i)$ to a final point $(f \equiv (x_f, y_f)$ then the corresponding change in is F given by

$$\Delta F = F_f - F_i = \int_i^f dF = \int_i^f (X\, dx + Y\, dy).$$

Note that since the difference on the left-hand side depends only on the initial and final points, the integral on the right-hand side can only depend on these points as well. In other words, the value of the integral is independent of the path taken in going from the initial to the final point. This is the distinguishing feature of an exact differential. Consider an integral taken around a closed circuit in the $x - y$ plane. In this case, the initial and final points correspond to the same point, so the difference $F_f - F_i$ is clearly zero. It follows that the integral of an exact differential over a closed circuit is always zero:

$$\oint dF \equiv 0.$$

Of course, not every infinitesimal quantity is an exact differential. Consider the infinitesimal object

$$đG \equiv X'(x,y)dx + Y'(x,y)dz,$$

where X' and Y're two general functions of x and y. It is easy to test whether or not an infinitesimal quantity is an exact differential. Consider the expression. It is clear that since $X = \partial F/\partial x$ and $Y = \partial F/\partial y$then

$$\frac{\partial X}{\partial y} = \frac{\partial Y}{\partial x} = \frac{\partial^2 F}{\partial x \partial y}.$$

Thus, if

$$\frac{\partial X'}{\partial y} \neq \frac{\partial Y'}{\partial x}$$

(as is assumed to be the case), then $đG$ cannot be an exact differential, and is instead termed an inexact differential. The special symbol $đ$ is used to denote an inexact differential. Consider the integral of $đG$ over some path in the x–y plane. In general, it is not true that

$$\int_i^f đG = \int_i^f (X'dx + Y'dy)$$

is independent of the path taken between the initial and final points. This is the distinguishing feature of an inexact differential. In particular, the integral of an inexact differential around a closed circuit is not necessarily zero, so

$$\oint đG \neq 0.$$

Consider, for the moment, the solution of

$$đG = 0,$$

which reduces to the ordinary differential equation

$$\frac{dy}{dx} = -\frac{X'}{Y'}.$$

Since the right-hand side is a known function of x and y, the above equation defines a definite direction (i.e., gradient) at each point in the x–y plane. The solution simply consists of drawing a system of curves in the x–y plane such that at any point the tangent to the curve is as specified in Eq. This defines a set of curves which can be written s(x, y) = c, where c is a labeling parameter. It follows that

$$\frac{d\sigma}{dx} \equiv \frac{\partial\sigma}{\partial x} + \frac{\partial\sigma}{\partial y}\frac{dy}{dx} = 0.$$

The elimination of dy/dx between Eqs. yields

$$Y'\frac{\partial\sigma}{\partial x} = X'\frac{\partial\sigma}{\partial y} = \frac{X'Y'}{\tau},$$

where $\tau(x, y)$is function of x and y. The above equation could equally well be written

$$X' = \frac{\partial\sigma}{\partial x}, \quad Y' = \tau\frac{\partial\sigma}{\partial y}.$$

Inserting Eq. gives

$$đG = \tau\left(\frac{\partial\sigma}{\partial x}dx + \frac{\partial\sigma}{\partial y}dy\right) = \tau\, d\sigma,$$

or

$$\frac{đG}{\tau} = d\sigma.$$

Thus, dividing the inexact differential $đG$ by τ yields the exact differential dσ. A factor τ which possesses this property is termed an integrating factor.

Since the above analysis is quite general, it is clear that an inexact differential involving two independent variables always admits of an integrating factor. Note, however, this is not generally the case for inexact differentials involving more than two variables.

After this mathematical excursion, let us return to physical situation of interest. The macrostate of a macroscopic system can be specified by the values of the external parameters (e.g., the volume) and the mean energy $\bar{E}$. This, in turn, fixes other parameters such as the mean pressure $\bar{p}$. Alternatively, we can specify the external parameters and the mean pressure, which fixes the mean energy. Quantities such as $d\bar{p}$ and $d\bar{E}$ are infinitesimal differences between well-defined quantities: i.e., they are exact differentials.

For example, $d\bar{E} = \bar{E}_f - \bar{E}_i$ is just the difference between the mean energy of the system in the final macrostate f and the initial macrostate i, in the limit where these two states are nearly the same. It follows that if the system is taken from an initial macrostate i to any final macrostate f the mean energy change is given by

$$\Delta\bar{E} = \bar{E}_f - \bar{E}_i = \int_i^f d\bar{E}.$$

However, since the mean energy is just a function of the macrostate under consideration, $\bar{E}_f$ and $\bar{E}_i$ depend only on the initial and final states, respectively. Thus, the integral $\int d\bar{E}$ depends only on the initial and final states, and not on the particular process used to get between them.

Consider, now, the infinitesimal work done by the system in going from some initial macrostate to some neighbouring final macrostate f. In general, $đW = \sum \bar{X}_\alpha \bar{d}x_\alpha$ is not the difference between two numbers referring to the properties of two neighbouring macrostates. Instead, it is merely an infinitesimal quantity characteristic of the process of going from state to state f. In other words, the work $đW$ is in general an inexact differential. The total work done by the system in going from any macrostate to some other macrostate f can be written as

$$W_{if} = \int_i^f đW_1$$

where the integral represents the sum of the infinitesimal amounts of work $đW$ performed at each stage of the process. In general, the value of the integral does depend on the particular process used in going from macrostate to macrostate f.

Recall that in going from macrostate i to macrostate f the change $\Delta\bar{E}$ does not depend on the process used whereas the work W, in general, does. Thus, it follows from the first law of thermodynamics, Eq. that the heat Q, in general, also depends on the process used. It follows that

$$đQ \equiv d\bar{E} + đW$$

is an inexact differential. However, by analogy with the mathematical example, there must exist some integrating factor, T, say, which converts the inexact differential $đQ$ into an exact differential. So,

$$\frac{đQ}{T} \equiv dS.$$

It will be interesting to find out what physical quantities correspond to the functions T and S.

Suppose that the system is thermally insulated, so that $Q = 0$. In this case, the first law of thermodynamics implies that

$$W_{if} = -\Delta\bar{E}.$$

Thus, in this special case, the work done depends only on the energy difference between in the initial and final states, and is independent of the process. In fact, when Clausius first formulated the first law in 1850 this is how he expressed it:

If a thermally isolated system is brought from some initial to some final state then the work done by the system is independent of the process used.

If the external parameters of the system are kept fixed, so that no work is done, then $đW = 0$, Eq. reduces to

$$đQ - d\bar{E},$$

and $đQ$ becomes an exact differential. The amount of heat Q absorbed in going from one macrostate to another depends only on the mean energy difference between them, and is independent of the process used to effect the change. In this situation, heat is a conserved quantity, and acts very much like the invisible indestructible fluid of Lavoisier's calorific theory.

FEELING AND SEEING TEMPERATURE CHANGES

Within some reasonable temperature range, we can get a rough idea how warm something is by touching it. But this can be unreliable—if you put one hand in cold water, one in hot, then plunge both of them into lukewarm water, one hand will tell you it's hot, the other will feel cold. For something too hot to touch, we can often get an impression of how hot it is by approaching and sensing the radiant heat. If the temperature increases enough, it begins to glow.

The problem with these subjective perceptions of heat is that they may not be the same for everybody. If our two hands can't agree on whether water is warm or cold, how likely is it that a group of people can set a uniform standard? We need to construct a device of some kind that responds to temperature in a simple, measurable way—we need a thermometer. The first step on the road to a thermometer was taken by one Philo of Byzantium, an engineer, in the second century BC. He took a hollow lead sphere connected with a tight seal to one end of a pipe, the other end of the pipe being under water in another vessel.

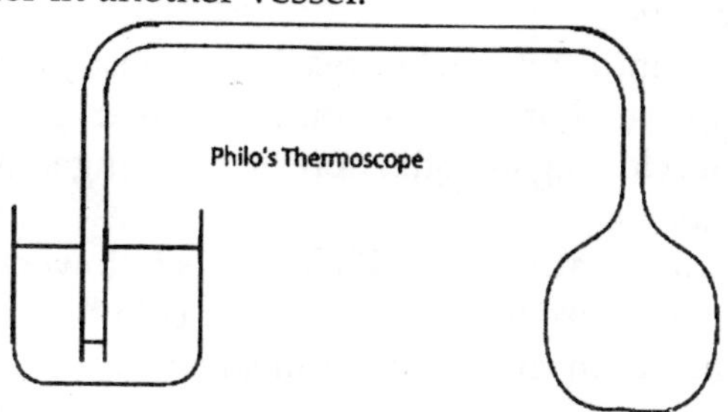

Fig. Philo's Thermoscope

To quote Philo: "...if you expose the sphere to the sun, part of the air enclosed in the tube will pass out when the sphere becomes hot. This will be evident because the air will descend from the tube into the water, agitating it and producing a succession of bubbles.

Now if the sphere is put back in the shade, that is, where the sun's rays do not reach it, the water will rise and pass through the tube ..."

"No matter how many times you repeat the operation, the same thing will happen.

In fact, if you heat the sphere with fire, or even if you pour hot water over it, the result will be the same."

Notice that Philo did what a real investigative scientist should do—he checked that the experiment was reproducible, and he established that the air's expansion was in response to heat being applied to the sphere, and was independent of the source of the heat.

This expansion of air on heating became widely known in classical times, and was used in various dramatic devices. For example, Hero of Alexandria describes a small temple where a fire on the altar causes the doors to open.

The altar is a large airtight box, with a pipe leading from it to another enclosed container filled with water. When the fire is set on top of the altar, the air in the box heats up and expands into a second container which is filled with water. This water is forced out through an overflow pipe into a bucket hung on a rope attached to the door hinges in such a way that as the bucket fills with water, it drops, turns the hinges, and opens the doors.

The pipe into this bucket reaches almost to the bottom, so that when the altar fire goes out, the water is sucked back and the doors close again. Still, none of these ingenious devices is a thermometer.

There was no attempt by Philo or his followers to make a quantitative measurement of how hot or cold the sphere was. And the "meter" in thermometer means measurement.

THE FIRST THERMOMETER

Galileo claimed to have invented the first thermometer. Well, actually, he called it a thermoscope, but he did try to measure "degrees of heat and cold" according to a colleague, and that qualifies it as a thermometer. Galileo used an inverted narrow-necked bulb with a tubular neck, like a hen's egg with a long glass tube attached at the tip.

He first heated the bulb with his hands then immediately put it into water. He recorded that the water rose in the bulb the height of "one palm". Later, either Galileo or his colleague

Santorio Santorio put a paper scale next to the tube to read off changes in the water level. This definitely made it a thermometer, but who thought of it first isn't clear (they argued about it). And, in fact, this thermometer had problems.

Question: what problems? If you occasionally top up the water, why shouldn't this thermometer be good for recording daily changes in temperature?

Answer: because it's also a barometer! But—Galileo didn't know about the atmospheric pressure.

Torricelli, one of Galileo's pupils, was the first to realize, shortly after Galileo died, that the real driving force in suction was external atmospheric pressure, a satisfying mechanical explanation in contrast to the philosophical "nature abhors a vacuum". In the 1640's, Pascal pointed out that the variability of atmospheric pressure rendered the air thermometer untrustworthy.

Liquid-in-glass thermometers were used from the 1630's, and they were of course insensitive to barometric pressure. Meteorological records were kept from this time, but there was no real uniformity of temperature measurement until Fahrenheit, almost a hundred years later.

NEWTON'S ANONYMOUS TABLE OF TEMPERATURES

Presumably he felt that this project lacked the timeless significance of some of his other achievements. Taking the freezing point of water as zero, Newton found the temperature of boiling water to be almost three times that of the human body, melting lead eight times as great (actually 327C, whereas 8×37=296, so this is pretty good!) but for higher temperatures, such as that of a wood fire, he underestimated considerably. He used a linseed oil liquid in glass thermometer up to the melting point of tin (232°C). Newton tried to estimate the higher temperatures indirectly. He heated up a piece of iron in a fire, then let it cool in a steady breeze.

He found that, at least at the lower temperatures where

he could cross check with his thermometer, the temperature dropped in a geometric progression, that is, if it took five minutes to drop from 80° above air temperature to 40° above air temperature, it took another five minutes to drop to 20° above air, another five to drop to 10° above, and so on. He then assumed this same pattern of temperature drop was true at the high temperatures beyond the reach of his thermometer, and so estimated the temperature of the fire and of iron glowing red hot. This wasn't very accurate—he (under)estimated the temperature of the fire to be about 600°C.

FAHRENHEIT'S EXCELLENT THERMOMETER

The first really good thermometer, using mercury expanding from a bulb into a capillary tube, was made by Fahrenheit in the early 1720's. He got the idea of using mercury from a colleague's comment that one should correct a barometer reading to allow for the variation of the density of mercury with temperature.

The point that has to be borne in mind in constructing thermometers, and defining temperature scales, is that not all liquids expand at uniform rates on heating—water, for example, at first contracts on heating from its freezing point, then begins to expand at around forty degrees Fahrenheit, so a water thermometer wouldn't be very helpful on a cold day.

It is also not easy to manufacture a uniform cross section capillary tube, but Fahrenheit managed to do it, and demonstrated his success by showing his thermometers agreed with each other over a whole range of temperatures. Fortunately, it turns out that mercury is well behaved in that the temperature scale defined by taking its expansion to be uniform coincides very closely with the true temperature scale.

Amontons' Air Thermometer

A little earlier (1702) Amontons introduced an air pressure thermometer. He established that if air at atmospheric pressure at the freezing point of water is enclosed then heated to the boiling point of water, but meanwhile kept at constant volume by increasing the pressure on it, the pressure goes up by about 10 inches of mercury.

He also discovered that if he compressed the air in the first place, so that it was at a pressure of sixty inches of mercury at the temperature of melting ice, then if he raised its temperature to that of boiling water, at the same time adding mercury to the column to keep the volume of air constant, the pressure increased by 20 inches of mercury. In other words, he found that for a fixed amount of air kept in a container at constant volume, the pressure increased with temperature by about 33% from freezing to boiling, that percentage being independent of the initial pressure.

THE ZEROTH LAW OF THERMODYNAMICS

Once the thermometer came to be widely used, more precise observations of temperature and heat flow became possible. Joseph Black, a professor at the University of Edinburgh in the 1700's, noticed that a collection of objects at different temperatures, if brought together, will all eventually reach the same temperature.

"By the use of these instruments [thermometers] we have learned, that if we take 1000, or more, different kinds of matter, such as metals, stones, salts, woods, cork, feathers, wool, water and a variety of other fluids, although they be all at first of different heats, let them be placed together in a room without a fire, and into which the sun does not shine, the heat will be communicated from the hotter of these bodies to the colder, during some hours, perhaps, or the course of a day, at the end of which time, if we apply a thermometer to all of them in succession, it will point to precisely the same degree."

We say nowadays that bodies in "thermal contact" eventually come into "thermal equilibrium"—which means they finally attain the same temperature, after which no further heat flow takes place. This is equivalent to:

If two objects are in thermal equilibrium with a third, then they are in thermal equilibrium with each other.

The "third body" in a practical situation is just the thermometer. It's perhaps worth pointing out that this trivial sounding statement certainly wasn't obvious before the invention of the thermometer. With only the sense of touch to go on, few people would agree that a piece of wool and a bar of metal, both at 0°C, were at the same temperature. "flow of heat" that takes place between bodies as they move towards thermal equilibrium? For example, suppose I reproduce one of Fahrenheit's experiments, by taking 100 ccs of water at 100°F, and 100ccs at 150°F, and mix them together in an insulated jug so little heat escapes. What is the final temperature of the mix?

Of course, it's close to 125°F—not surprising, but it does tell us something! It tells us that the amount of heat required to raise the temperature of 100 cc of water from 100°F to 125°F is exactly the same as the amount needed to raise it from 125°F to 150°F. A series of such experiments established that it always took the same amount of heat to raise the temperature of 1 cc of water by one degree, independent of the initial temperature of the water (provided it's between the freezing point and the boiling point). This makes it possible to define a unit of heat. Perhaps unfairly to Fahrenheit, 1 calorie is the heat required to raise the temperature of 1 gram of water by 1 degree Celsius.

(Celsius also lived in the early 1700's. His scale has the freezing point of water as 0°C, the boiling point as 100°C. Fahrenheit's scale is no longer used in science, but lives on in engineering in the US, and in the British Thermal Unit, which is the heat required to raise the temperature of one pound of water by 1°F).

SPECIFIC HEATS AND CALORIMETRY

The specific heat of a substance is the heat required in calories to raise the temperature of 1 gram by 1 degree Celsius. As Fahrenheit continues his measurements of heat flow, it quickly became evident that for different materials, the amount of heat needed to raise the temperature of one gram by one degree could be quite different. It was natural to measure specific heats relative to that of water, the simplest and most readily available substance for calorimetric experiments.

There were some surprises. For example, it had been widely thought before the measurements were made, that one cc of Mercury, being a lot heavier than one cc of water, would take more heat to raise its temperature by one degree. This proved not to be the case—Fahrenheit himself made the measurement. In an insulating container, called a "calorimeter" he added 100ccs of water at 100°F to 100ccs of mercury at 150°F, and stirred so they quickly reached thermal equilibrium.

Question: what do you think the final temperature was?

Answer: The final temperature was, surprisingly, about 120°F. 100 cc of water evidently "contained more heat" than 100 cc of mercury, despite the large difference in weight.

The technique, called calorimetry, was widely used to find specific heats of many different substances, and at first no clear pattern emerged. It was puzzling that the specific heat of mercury was so low compared with water. As the experiments progressed, it gradually became evident that heavier substances, paradoxically, had lower specific heats.

Meanwhile, this quantitative approach to scientific observation had spread to chemistry. Towards the end of the 1700's, Lavoisier weighed chemicals involved in reactions before and after the reaction.

This involved weighing the gases involved, so had to be carried out in closed containers, so that, for example, the weight of oxygen used and the carbon dioxide, etc., produced would accounted for in studying combustion. The big discovery was that mass was neither created nor destroyed. This had not been realized before because no one had weighed the gases involved. It made the atomic theory suddenly more plausible, with the idea that maybe chemical reactions were just rearrangements of atoms into different combinations.

Lavoisier also clarified the concept of an element, an idea that was taken up in about 1800 by John Dalton, who argues that a given compound consisted of identical molecules, made up of elementary atoms in the same proportion, such as H2O (although that was thought initially to be HO). This explained why, when substances reacted chemically, such as the burning of hydrogen to form water, it took exactly eight grams of oxygen for each gram of hydrogen. (produce H2O2 under the right conditions, with exactly sixteen grams of oxygen to one of hydrogen, but the simple ratios of amounts of oxygen needed for the two reactions were simply explained by different molecular structures, and made the atomic hypothesis even more plausible.)

Much effort was expended carefully weighing the constituents in many chemical reactions, and constructing diagrams of the molecules. The important result of all this work was that it became possible to list the relative weights of the atoms involved.

For example, the data on H2O and H2O2 led to the conclusion that an oxygen atom weighed sixteen times the weight of a hydrogen atom.

It must be emphasized, though, that these results gave no clue as to the actual weights of atoms! All that was known was that atoms were too small to see in the best microscopes. Nevertheless, knowing the relative weights of some atoms in 1820 led to an important discovery.

Two professors in France, Dulong and Petit, found that for a whole series of elements the product of atomic weight and specific heat was the same.

Element	*Specific Heat (cals/gm/degree C)*	*Relative weights of the atoms*	*Product of relative atomic weight and specific heat*
Lead	0.0293	12.95	0.3794
Tin	0.0514	7.35	0.3779
Zinc	0.0927	4.03	0.3736
Sulphur	0.188	2.011	0.378

The significance of this, as they pointed out, was that the "specific heat", or heat capacity, of each atom was the same—a piece of lead and a piece of zinc having the same number of atoms would have the same heat capacity. So heavier atoms absorbed no more heat than lighter atoms for a given rise in temperature.

This partially explained why mercury had such a

surprisingly low heat capacity. Of course, having no idea how big the atoms might be, they could go no further. And, indeed, many of their colleagues didn't believe in atoms anyway, so it was hard to convince them of the significance of this discovery.

LATENT HEAT

One of Black's experiments was to set a pan of water on a steady fire and observe the temperature as a function of time. He found it steadily increased, reflecting the supply of heat from the fire, until the water began to boil, whereupon the temperature stayed the same for a long time. The steam coming off was at the same (boiling) temperature as the water. So what was happening to the heat being supplied? Black correctly concluded that heat needed to be supplied to change water from its liquid state to its gaseous state, that is, to steam. In fact, a lot of heat had to be supplied: 540 calories per gram, as opposed to the mere 100 calories per gram needed to bring it from the freezing temperature to boiling.

He also discovered that it took 80 calories per gram to melt ice into water, with no rise in temperature. This heat is released when the water freezes back to ice, so it is somehow "hidden" in the water. He called it latent heat, meaning hidden heat.

THE GAS LAW

Coefficients of Expansion

Almost all materials expand on heating—the most famous exception being water, which contracts as it is warmed from 0 degrees Celsius to 4 degrees. This is actually a good thing, because as freezing weather sets in, the coldest water, which is about to freeze, is less dense than slightly warmer water, so rises to the top of a lake and the ice begins to form there. For almost all other liquids, solidification on cooling begins at the bottom of the container. So, since water behaves in this weird way, ice skating is possible! Also, as a matter of fact, life in lakes is possible—the ice layer that forms insulates the rest of the lake water from very cold air, so fish can make it through the winter.

LINEAR EXPANSION

The coefficient of linear expansion α of a given material, for example a bar of copper, at a given temperature is defined as the fractional increase in length that takes place on heating through one degree:

$$L \to L + \Delta L = (1+\alpha)\,L \; when T \to T + 1°C.$$

Of course, α might vary with temperature (it does for water, as we just mentioned) but in fact for most materials it stays close to constant over wide temperature ranges.

For copper, $\alpha = 17 \times 10^{-6}$

VOLUME EXPANSION

For liquids and gases, the natural measure of expansion is the coefficient of volume expansion, β.

$$V \to V + \Delta V = (1+\beta)V \text{ when } T \to T + 1°C.$$

Of course, on heating a bar of copper, clearly the volume as well as the length increases—the bar expands by an equal fraction in all directions (this could be experimentally verified, or you could just imagine a cube of copper, in which case all directions look the same).

The volume of a cube of copper of side L is V = L3. Suppose we heat it through one degree. Putting together the definitions of α, β above,

$$V \to (1+\beta)V, L \to (1+\alpha)L, \; L^3 \to (1+\alpha^3)\,L^3 or V \to (1+a)^3\,V.$$

So $(1 + \beta) = (1 + \alpha)^3$. But remember α is very, very small—so even though $(1 + \alpha)^3 = 1 + 3\alpha + \alpha^2 + \alpha^3$, the last two terms are completely negligible so to a fantastically good approximation:

$$\beta = 3\alpha$$

The coefficient of volume expansion is just three times the coefficient of linear expansion.

GAS PRESSURE INCREASE WITH TEMPERATURE

In 1702, Amontons discovered a linear increase of P with

T for air, and found P to increase about 33% from the freezing point of water to the boiling point of water. That is to say, he discovered that if a container of air were to be sealed at 0°C, at ordinary atmospheric pressure of 15 pounds per square inch, and then heated to 100°C but kept at the same volume, the air would now exert a pressure of about 20 pounds per square inch on the sides of the container. Remarkably, Amontons discovered, if the gas were initially at a pressure of thirty pounds per square inch at 0°C, on heating to 100°C the pressure would go to about 40 pounds per square inch—so the percentage increase in pressure was the same for any initial pressure: on heating through 100°C, the pressure would always increase by about 33%.

Furthermore, the result turned out to be the same for different gases! Finding a Natural Temperature Scale.In class, we plotted air pressure as a function of temperature for a fixed volume of air, by making several measurements as the air was slowly heated (to give it a chance to all be at the same temperature at each stage).

We found a straight line. On the graph, we extended the line backwards, to see how the pressure would presumably drop on cooling the air.

We found the remarkable prediction that the pressure should drop to zero at a temperature of about –273°C. In fact, if we'd done the cooling experiment, we would have found that air doesn't actually follow the line all the way down, but condenses to a liquid at around –200°C.

However, helium gas stays a gas almost to –270°C, and follows the line closely. The physics of gases, and the interpretation of this, much more fully in a couple of lectures. For now, the important point is that this suggests a much more natural temperature scale than the Celsius one: we should take -273°C as the zero of temperature! For one thing, if we do that, the pressure/temperature relationship for a gas becomes beautifully simple:

$$P \propto T.$$

This temperature scale, in which the degrees have the same size as in Celsius, is called the Kelvin or absolute scale.

Temperatures are written 300K. To get from Celsius to Kelvin, just add 273.15.

AN IDEAL GAS

Physicists at this point introduce the concept of an "Ideal Gas". This is like the idea of a frictionless surface: it doesn't exist in nature, but it is a very handy approximation to some real systems, and makes problems much easier to handle mathematically. The ideal gas is one for which $P \propto T$ for all temperatures, so helium is close to ideal over a very wide range, and air is close to ideal at ordinary atmospheric temperatures and above.

THE GAS LAW

We say earlier in the course that for a gas at constant temperature PV = constant (Boyle's Law). Now at constant volume,

$$P \propto T$$

We can put these together in one equation to find a relationship between pressure, volume and temperature:

$$PV = CT$$

where C is a constant. Notice, by the way, that we can immediately conclude that at fixed pressure, $V \propto T$, this is called Charles' Law. But what is C? Obviously, it depends on how much gas we have—double the amount of gas, keeping the pressure and temperature the same, and the volume will be doubled, so C will be doubled.

But notice that C will not depend on what gas we are talking about: if we have two separate one-liter containers, one filled with hydrogen, the other with oxygen, both at atmospheric pressure, and both at the same temperature, then C will be the same for both of them. One might conclude from this that C should be defined for one liter of gas at a specified temperature and pressure, such as 0°C and 1 atmosphere, and that could be a consistent scheme. It might seem more natural, though, to specify a particular mass of gas, since then we wouldn't have to specify a particular temperature and pressure in the definition of C.

But that idea brings up a further problem: one gram of oxygen takes up a lot less room than one gram of hydrogen. Since we've just seen that choosing the same volume for the two gases gives the same constant C for the two gases, evidently taking the same mass of the two gases will give different C's.

AVOGADRO'S HYPOTHESIS

The resolution to this difficulty is based on a remarkable discovery the chemists made two hundred years or so ago: they found that one liter of nitrogen could react with exactly one liter of oxygen to produce exactly two liters of NO, nitrous oxide, all volume measurements being at the same temperature and pressure. Further, one liter of oxygen combined with two liters of hydrogen to produce two liters of steam.

These simple ratios of interacting gases could be understood if one imagined the atoms combining to form molecules, and made the further assumption, known as Avogadro's Hypothesis :

Equal volumes of gases at the same temperature and pressure contain the same number of molecules

One could then understand the simple volume results by assuming the gases were made of diatomic molecules, H2, N2, O_2 and the chemical reactions were just molecular recombinations given by the equations $N_2 + O_2 = 2NO$, $2H_2 + O_2 = 2H_2O$, etc. Of course, in 1811 Avogadro didn't have the slightest idea what this number of molecules was for, say, one liter, and nobody else did either, for another fifty years. So no-one knew what an atom or molecule weighed, but assuming that chemical reactions were atoms combining into molecules, or rearranging from one molecular pairing or grouping to another, they could figure out the relative weights of atoms, such as an oxygen atom had mass 16 times that of a hydrogen atom—even though they had no idea how big these masses were!

This observation led to defining the natural mass of a gas for setting the value of the constant C in the gas law to be a "mole" of gas: hydrogen was known to be H_2 molecules, so a mole of hydrogen was 2 grams, oxygen was O_2, so a mole of oxygen was 32 grams, and so on. With this definition, a mole of oxygen contains the same number of molecules as a mole of hydrogen: so at the same temperature and pressure, they will occupy the same volume. At 0°C, and atmospheric pressure, the volume is 22.4 liters. So, for one mole of a gas (for example, two grams of hydrogen), we set the constant C equal to R, known as the universal gas constant, and equal to 8.3 J/(mol.K) and PV = RT.

For n moles of a gas, such as 2n grams of hydrogen, the law is:

$$PV = nRT$$

and this is the standard form of the Gas Law.

By the late 1700's, the experiments of Fahrenheit, Black and others had established a systematic, quantitative way of measuring temperatures, heat flows and heat capacities—but this didn't really throw any new light on just what was flowing.

This was a time when the study of electricity was all the rage, led in America by Benjamin Franklin, who had suggested in 1747 that electricity was one (invisible) fluid (it had previously been suggested that there were two fluids, corresponding to the two kinds of electrical charging observed).

Chapter 3

Statistical Thermodynamics

Let us briefly review the material which we have covered so far in this course. We started off by studying the mathematics of probability. We then used probabilistic reasoning to analyze the dynamics of many particle systems, a subject area known as statistical mechanics. Next, we explored the physics of heat and work, the study of which is termed thermodynamics. The final step in our investigation is to combine statistical mechanics with thermodynamics: in other words, to investigate heat and work via statistical arguments. This discipline is called statistical thermodynamics, and forms the central subject matter of this course.

THERMAL INTERACTION BETWEEN MACROSYSTEMS

Let us begin our investigation of statistical thermodynamics by examining a purely thermal interaction between two macroscopic systems, A and A', from a microscopic point of view. Suppose that the energies of these two systems are E and E', respectively. The external parameters are held fixed, so that A and A' cannot do work on one another.

However, we assume that the systems are free to exchange heat energy (i.e., they are in thermal contact). It is convenient to divide the energy scale into small subdivisions of width δE. The number of microstates of A consistent with a macrostate in which the energy lies in the range E to $E + \delta E$ is denoted $\Omega(E)$. Likewise, the number of microstates of consistent with a macrostate in which the energy lies between E' and $E' + \delta E$ is denoted $\Omega'(E')$.

The combined system $A^{(0)} = A + A'$ is assumed to be isolated (i.e., it neither does work on nor exchanges heat with its surroundings). It follows from the first law of thermodynamics that the total energy $E^{(0)}$ is constant. When speaking of thermal contact between two distinct systems, we usually assume that the mutual interaction is sufficiently weak for the energies to be additive. Thus,

$$E + E = E^{(0)} = \text{constant}$$

Of course, in the limit of zero interaction the energies are strictly additive. However, a small residual interaction is always required to enable the two systems to exchange heat energy and, thereby, eventually reach thermal equilibrium. In fact, if the interaction between A and A' is too strong for the energies to be additive then it makes little sense to consider each system in isolation, since the presence of one system clearly strongly perturbs the other, and vice versa. In this case, the smallest system which can realistically be examined in isolation is $A^{(0)}$.

According to Eq. if the energy of A lies in the range E to $E + \delta E$ then the energy of A' must lie between $E^{(0)} - E - \delta E$ and $E^{(0)} - E$. Thus, the number of microstates accessible to each system is given by $\Omega'(E)$ and $(E^{(0)} - E)$, respectively. Since every possible state of A can be combined with every possible state of A' to form a distinct microstate, the total number of distinct states accessible to $A^{(0)}$ when the energy of A lies in the range E to $E + \delta E$ is $\Omega^{(0)} = \Omega(E)\, \Omega'(E^{(0)} - E)$.

Consider an ensemble of pairs of thermally interacting systems, A and A', which are left undisturbed for many relaxation times so that they can attain thermal equilibrium. The principle of equal a priori probabilities is applicable to this situation.

According to this principle, the probability of occurrence of a given macrostate is proportional to the number of accessible microstates, since all microstates are equally likely. Thus, the probability that the system A has an energy lying in the range E to $E + \delta E$can be written

$$P(E) = C\, \Omega(E)\, \Omega'\, (E^{(0)} - E),$$

where C is a constant which is independent of E.

The typical variation of the number of accessible states with energy is of the form

$$\Omega \propto E^f,$$

where f is the number of degrees of freedom. For a macroscopic system f is an exceedingly large number. It follows that the probability $P(E)$in Eq. is the product of an extremely rapidly increasing function of E and an extremely rapidly decreasing function of E. Hence, we would expect the probability to exhibit a very pronounced maximum at some particular value of the energy.

Let us Taylor expand the logarithm of $P(E)$ in the vicinity of its maximum value, which is assumed to occur at $E = \tilde{E}$. We expand the relatively slowly varying logarithm, rather than the function itself, because the latter varies so rapidly with the energy that the radius of convergence of its Taylor expansion is too small for this expansion to be of any practical use. The expansion of ln Ω (E) yields

$$\ln \Omega(E) = \ln \Omega(\tilde{E}) + \beta(\tilde{E})\eta - \frac{1}{2}\lambda(\tilde{E})\eta^2 + \cdots,$$

where

$$\eta = E - \tilde{E},$$

$$\beta = \frac{\partial \ln \Omega}{\partial E},$$

$$\lambda = \frac{\partial^2 \ln \Omega}{\partial E^2} = -\frac{\partial \beta}{\partial E}.$$

Now, since $E' = E^{(0)} - E$, we have

$$E' - \tilde{E}' = -(E - \tilde{E}) = -\eta.$$

It follows that

$$\ln \Omega'(E') = \ln \Omega'(\tilde{E}') + \beta'(\tilde{E}')(-\eta) - \frac{1}{2}\lambda'(\tilde{E}')(-\eta)^2 + \cdots,$$

where $\beta'$and $\lambda'$are defined in an analogous manner to the parameters β and λ. Equations can be combined to give

$$\ln[\Omega(E)\Omega'(E')] = \ln[\Omega(\tilde{E})\Omega'(\tilde{E}')] + [\beta(\tilde{E}) - \beta'(\tilde{E}')]\eta - \frac{1}{2}[\lambda(\tilde{E}) + \lambda'(\tilde{E}')]\eta^2 + \cdots.$$

At the maximum of ln $[\Omega(E)\Omega'(E')]$ the linear term in the Taylor expansion must vanish, so

$$\beta(\tilde{E}) = \beta(\tilde{E}'),$$

which enables us to determine $\tilde{E}$. It follows that

$$\ln P(E) = \ln P(\tilde{E}) - \frac{1}{2}\lambda_0 \eta^2,$$

or

$$P(E) = P(\tilde{E}) \exp\left[-\frac{1}{2}\lambda_0 (E - \tilde{E})^2\right],$$

where

$$\lambda_0 = \lambda(\tilde{E}) + \lambda'(\tilde{E}').$$

Now, the parameter λ_0 must be positive, otherwise the probability $P(E)$ does not exhibit a pronounced maximum value: i.e., the combined system $A^{(0)}$ does not possess a well-defined equilibrium state as, physically, we know it must. It is clear that $\lambda(\tilde{E})$ must also be positive, since we could always choose for A' a system with a negligible contribution to λ_0, in which case the constraint $\lambda_0 > 0$ would effectively correspond to $\lambda(\tilde{E}) > 0$. [A similar argument can be used to show that $\lambda'(\tilde{E}')$ must be positive.] The same conclusion also follows from the estimate $\Omega \propto E^f$, which implies that

$$\lambda(\tilde{E}) \sim \frac{f}{\tilde{E}^2} > 0 .$$

According to Eq. the probability distribution function $P(E)$ is a Gaussian. This is hardly surprising, since the central limit theorem ensures that the probability distribution for any macroscopic variable, such as $\tilde{E}$, is Gaussian in nature. It follows that the mean value of E corresponds to the situation of maximum probability (i.e., the peak of the Gaussian curve), so that

$$\overline{E} = \tilde{E}.$$

The standard deviation of the distribution is

$$\Delta * E = \lambda_0^{-1/2} \sim \frac{\overline{E}}{\sqrt{f}},$$

where use has been made of Eq. (assuming that system makes the dominant contribution to λ_0). It follows that the fractional width of the probability distribution function is given by

$$\frac{\Delta^* E}{\overline{E}} \sim \frac{1}{\sqrt{f}}.$$

Hence, if A contains 1 mole of particles then $f \sim N_A \simeq 10^{24}$ and $\Delta^* E/\overline{E} \sim 10^{-12}$. Clearly, the probability distribution for E has an exceedingly sharp maximum. Experimental measurements of this energy will almost always yield the mean value, and the underlying statistical nature of the distribution may not be apparent.

TEMPERATURE

Suppose that the systems A and A' are initially thermally isolated from one another, with respective energies E_i and E_i'. (Since the energy of an isolated system cannot fluctuate, we do not have to bother with mean energies here.) If the two systems are subsequently placed in thermal contact, so that they are free to exchange heat energy, then, in general, the resulting state is an extremely improbable one [i.e., $P(E_i)$is much less than the peak probability]. The configuration will, therefore, tend to change in time until the two systems attain final mean energies $\overline{E}_f$ and $\overline{E}_f'$ which are such that

$$\beta_f = \beta_f',$$

where $\beta_f \equiv \beta(\overline{E}_f)$ and $\beta_f' \equiv \beta'(\overline{E}_f')$. This corresponds to the state of maximum probability. In the special case where the initial energies, E_i and E_i', lie very close to the final mean energies, $\overline{E}_f$ and $\overline{E}_f'$, respectively, there is no change in the two systems when they are brought into thermal contact, since the initial state already corresponds to a state of maximum probability.

It follows from energy conservation that

$$\overline{E}_f + \overline{E}_f' = E_i + E_i'.$$

The mean energy change in each system is simply the net heat absorbed, so that

$$Q \equiv \overline{E}_f - E_i,$$

$$Q' \equiv \overline{E}'_f - E'_i.$$

The conservation of energy then reduces to

$$Q + Q' = 0:$$

i.e., the heat given off by one system is equal to the heat absorbed by the other (in our notation absorbed heat is positive and emitted heat is negative).

It is clear that if the systems A and A' are suddenly brought into thermal contact then they will only exchange heat and evolve towards a new equilibrium state if the final state is more probable than the initial one. In other words, if

$$P(\overline{E}_f) > (E_i),$$

or

$$\ln P(\overline{E}_f) > \ln P(E_i),$$

since the logarithm is a monotonic function. The above inequality can be written

$$\ln \Omega(\overline{E}_f) + \ln \Omega'(\overline{E}'_i) > \ln \Omega(E_i) + \ln \Omega'(E'_i),$$

with the aid of Eq. Taylor expansion to first order yields

$$\frac{\partial \ln \Omega(E_i)}{\partial E}(\overline{E}_f - E_i) + \frac{\partial \ln \Omega'(E'_i)}{\partial E'}(\overline{E}'_f - E'_i) > 0,$$

which finally gives

$$(\beta_i - \beta'_i)\,Q > 0,$$

where $\beta_i \equiv \beta(E_i)$, $\beta'_i \equiv \beta'(E'_i)$, and use has been made of Eq.

It is clear, from the above, that the parameter β, defined

$$\beta = \frac{\partial \ln \Omega}{\partial E},$$

has the following properties:

If two systems separately in equilibrium have the same value of β then the systems will remain in equilibrium when brought into thermal contact with one another.

If two systems separately in equilibrium have different values of β then the systems will not remain in equilibrium when brought into thermal contact with one another. Instead, the system with the higher value of β will absorb heat from the other system until the two β values are the same.

Incidentally, a partial derivative is used in Eq. because in a purely thermal interaction the external parameters of the system are held constant whilst the energy changes.

Let us define the dimensionless parameter T, such that

$$\frac{1}{kT} \equiv \beta \equiv \frac{\partial \ln \Omega}{\partial E},$$

where k is a positive constant having the dimensions of energy. The parameter T is termed the thermodynamic temperature, and controls heat flow in much the same manner as a conventional temperature. Thus, if two isolated systems in equilibrium possess the same thermodynamic temperature then they will remain in equilibrium when brought into thermal contact.

However, if the two systems have different thermodynamic temperatures then heat will flow from the system with the higher temperature (i.e., the "hotter" system) to the system with the lower temperature until the temperatures of the two systems are the same. In addition, suppose that we have three systems A, B, and C. We know that if A and B remain in equilibrium when brought into thermal contact then their temperatures are the same, so that $T_A = T_B$. Similarly, if B and C remain in equilibrium when brought into thermal contact, then $T_B = T_C$. But, we can then conclude that $T_A = T_C$, so systems A and C will also remain in equilibrium when brought into thermal contact. Thus, we arrive at the following statement, which is sometimes called the zeroth law of thermodynamics:

If two systems are separately in thermal equilibrium with a third system then they must also be in thermal equilibrium with one another.

The thermodynamic temperature of a macroscopic body, as defined in Eq. depends only on the rate of change of the number of accessible microstates with the total energy. Thus, it is possible to define a thermodynamic temperature for systems with radically different microscopic structures (e.g., matter and radiation). The thermodynamic, or absolute, scale of temperature is measured in degrees kelvin. The parameter k is chosen to make this temperature scale accord as much as possible with more conventional temperature scales. The choice

$$k = 1.381 \times 10^{-23} \text{ joules/kelvin},$$

ensures that there are 100 degrees kelvin between the freezing and boiling points of water at atmospheric pressure (the two temperatures are 273.15 and 373.15 degrees kelvin, respectively). The above number is known as the Boltzmann constant. In fact, the Boltzmann constant is fixed by international convention so as to make the triple point of water (i.e., the unique temperature at which the three phases of water co-exist in thermal equilibrium) exactly 273.16°K. Note that the zero of the thermodynamic scale, the so called absolute zero of temperature, does not correspond to the freezing point of water, but to some far more physically significant temperature which.

The familiar $\Omega \propto E^f$ scaling for translational degrees of freedom yields

$$kT \sim \frac{\bar{E}}{f},$$

using Eq. so kT is a rough measure of the mean energy associated with each degree of freedom in the system. In fact, for a classical system (i.e., one in which quantum effects are unimportant) it is possible to show that the mean energy $(1/2)kT$ associated with each degree of freedom is exactly . This result, which is known as the equipartition theorem.

The absolute temperature T is usually positive, since $\Omega(E)$ is ordinarily a very rapidly increasing function of energy. In fact, this is the case for all conventional systems where the kinetic energy of the particles is taken into account, because there is no upper bound on the possible energy of the system, and $\Omega(E)$ consequently increases roughly like E^f. It is, however, possible to envisage a situation in which we ignore the translational degrees of freedom of a system, and concentrate only on its spin degrees of freedom. In this case, there is an upper bound to the possible energy of the system (i.e., all spins lined up anti-parallel to an applied magnetic field). Consequently, the total number of states available to the system is finite. In this situation, the density of spin states $\Omega_{spin}(E)$ first increases with increasing energy, as in conventional systems, but then reaches a maximum and decreases again. Thus, it is possible to get absolute spin temperatures which are negative, as well as positive.

In Lavoisier's calorific theory, the basic mechanism which forces heat to flow from hot to cold bodies is the supposed mutual repulsion of the constituent particles of calorific fluid. In statistical mechanics, the explanation is far less contrived. Heat flow occurs because statistical systems tend to evolve towards their most probable states, subject to the imposed physical constraints.

When two bodies at different temperatures are suddenly placed in thermal contact, the initial state corresponds to a spectacularly improbable state of the overall system. For systems containing of order 1 mole of particles, the only reasonably probable final equilibrium states are such that the two bodies differ in temperature by less than 1 part in 10^{12}. The evolution of the system towards these final states (i.e., towards thermal equilibrium) is effectively driven by probability.

Mechanical Interaction between Macrosystems

Let us now examine a purely mechanical interaction between macrostates, where one or more of the external parameters is modified, but there is no exchange of heat energy. Consider, for the sake of simplicity, a situation where only one external parameter x of the system is free to vary. In general, the number of microstates accessible to the system when the overall energy lies between E and $E + \delta E$ depends on the particular value of x, so we can write $\Omega \equiv \Omega(E, x)$.

When x is changed by the amount dx, the energy $E_r(x)$ of a given microstate r changes by $(\partial E_r/\partial x)dx$. The number of states $\sigma(E, x)$ whose energy is changed from a value less than E to a value greater than E when the parameter changes from x to $x + dx$ is given by the number of microstates per unit energy range multiplied by the average shift in energy of the microstates. Hence,

$$\sigma(E,x) = \frac{\Omega(E,x)}{\delta E}\overline{\frac{\partial E_r}{\partial x}}dx,$$

where the mean value of $\partial E_r/\partial x$is taken over all accessible microstates (i.e., all states where the energy lies between E and $E + \delta E$ and the external parameter takes the value). The above equation can also be written

$$s(E, x) = -\frac{\Omega(E,x)}{\delta E}\bar{X}\ dx,$$

where

$$\bar{X}(E,x) = -\overline{\frac{\partial E_r}{\partial x}}$$

is the mean generalized force conjugate to the external parameter x.

Consider the total number of microstates between E and $E + \delta E$. When the external parameter changes from to $x + dx$, the number of states in this energy range changes by $(\partial\Omega/\partial x)\ dx$. This change is due to the difference between the number of states which enter the range because their energy is changed from a value less than E to one greater than E and the number which leave because their energy is changed from a value less than $E + \delta E$ to one greater than $E + \delta E$. In symbols,

$$\frac{\partial\Omega\,(E,x)}{\partial x}dx = \sigma(E) - \sigma(E+\delta E) \simeq -\frac{\partial\sigma}{\partial E}\delta E,$$

which yields

$$\frac{\partial\Omega}{\partial x} = \frac{\partial(\Omega\bar{X})}{\partial E},$$

where use has been made of Eq. Dividing both sides by Ω gives

$$\frac{\partial\ln\Omega}{\partial x} = \frac{\partial\ln\Omega}{\partial E}\bar{X} + \frac{\partial\bar{X}}{\partial E}.$$

However, according to the usual estimate $\Omega \propto E^f$, the first term on the right-hand side is of order $(f/\bar{E})\,\bar{X}$, whereas the second term is only of order . Clearly, for a macroscopic system with many degrees of freedom, the second term is utterly negligible, so we have

$$\frac{\partial \ln \Omega}{\partial x} = \frac{\partial \ln \Omega}{\partial E}\bar{X} = \beta \bar{X}.$$

When there are several external parameters $x_1,\ldots, x_n$ so that $\Omega \equiv \Omega(E, x_1, \ldots, x_n)$, the above derivation is valid for each parameter taken in isolation. Thus,

$$\frac{\partial \ln \Omega}{\partial x_\alpha} = \beta \bar{X}_\alpha,$$

where $\bar{X}_\alpha$ is the mean generalized force conjugate to the parameter x_α.

General Interaction between Macrosystems

Consider two systems, A and A', which can interact by exchanging heat energy and doing work on one another. Let the system A' have energy E' and adjustable external parameters $x_1,\ldots,x_n$. Likewise, let the system $A^{(0)} = A + A'$have energy and adjustable external parameters $x_1',\ldots,x_n'$. The combined system $A^{(0)} = A + A'$is assumed to be isolated. It follows from the first law of thermodynamics that

$$E + E' = E^{(0)} = \text{constant}.$$

Thus, the energy E' of system A' is determined once the energy E of system A is given, and vice versa. In fact, E'could be regarded as a function of E. Furthermore, if the two systems can interact mechanically then, in general, the parameters x' are some function of the parameters x. As a simple example, if the two systems are separated by a movable partition in an enclosure of fixed volume $V^{(0)}$, then

$$V + V' = V^{(0)} = \text{constant},$$

where V and V' are the volumes of systems A and A', respectively.

The total number of microstates accessible to $A^{(0)}$ is clearly a function of E and the parameters x_α (where runs from 1 to n), so $\Omega^{(0)} \equiv \Omega^{(0)}(E, x_1,\ldots,x_n)$. We have already demonstrated that $\Omega^{(0)}$ exhibits a very pronounced maximum at one particular value of the energy $E = \tilde{E}$ when E is varied but the external parameters are held constant. This behaviour comes about because of the very strong,

$$\Omega \propto E^f,$$

increase in the number of accessible microstates of A(or A') with energy. The number of accessible microstates exhibits a similar strong increase with the volume, which is a typical external parameter, so that

$$\Omega \propto V^f.$$

It follows that the variation of $\Omega^{(0)}$ with a typical parameter x_α, when all the other parameters and the energy are held constant, also exhibits a very sharp maximum at some particular value $x_\alpha = \tilde{x}_\alpha$. The equilibrium situation corresponds to the configuration of maximum probability, in which virtually all systems $A^{(0)}$ in the ensemble have values of E and x_α very close to $\tilde{E}$ and $\tilde{x}_\alpha$. The mean values of these quantities are thus given by $\bar{E} = \tilde{E}$ and $\bar{x}_\alpha = \tilde{x}_\alpha$.

Consider a quasi-static process in which the system A is brought from an equilibrium state described by $\bar{E}$ and $\bar{x}_\alpha$ to an infinitesimally different equilibrium state described by and . Let us calculate the resultant change in the number of microstates accessible to A. Since $\Omega \equiv \Omega(E, x_1, ..., x_n)$, the change in ln Ω follows from standard mathematics:

$$d\ln\Omega = \frac{\partial \ln\Omega}{\partial E} d\bar{E} + \sum_{\alpha=1}^{n} \frac{\partial \ln\Omega}{\partial x_\alpha} d\bar{x}_\alpha .$$

However, we have previously demonstrated that

$$\beta = \frac{\partial \ln\Omega}{\partial E}, \beta\bar{X}_\alpha = \frac{\partial \ln\Omega}{\partial x_\alpha}$$

$$d\ln\Omega = \beta\left(d\bar{E} + \sum_{\alpha} \bar{X}_\alpha d\bar{x}_\alpha \right).$$

Note that the temperature β parameter and the mean conjugate forces $\bar{X}_\alpha$ are only well-defined for equilibrium states. This is why we are only considering quasi-static changes in which the two systems are always arbitrarily close to equilibrium.

Let us rewrite Eq. in terms of the thermodynamic temperature , T using the relation $\beta \equiv 1/kT$. We obtain

$$dS = \left(d\bar{E} + \sum_{\alpha} \bar{X}_{\alpha} d\bar{x}_{\alpha} \right) / T,$$

where

$$S = k \ln \Omega.$$

Equation is a differential relation which enables us to calculate the quantity S as a function of the mean energy $\bar{E}$ and the mean external parameters $\bar{x}_{\alpha}$, assuming that we can calculate the temperature T and mean conjugate forces $\bar{X}_{\alpha}$ for each equilibrium state. The function $S(\bar{E}, \bar{x}_{\alpha})$ is termed the entropy of system A. The word entropy is derived from the Greek en+trepien, which means "in change." The reason for this etymology will become apparent presently. It can be seen from Eq. that the entropy is merely a parameterization of the number of accessible microstates. Hence, according to statistical mechanics, is essentially a measure of the relative probability of a state characterized by values of the mean energy and mean external parameters $\bar{E}$ and $\bar{x}_{\alpha}$, respectively.

According to Eq. the net amount of work performed during a quasi-static change is given by

$$đW = \sum_{\alpha} \bar{X}_{\alpha} d\bar{x}_{\alpha}.$$

It follows from Eq. that

$$dS \frac{đW + đW}{T} = \frac{đQ}{T}.$$

Thus, the thermodynamic temperature T is the integrating factor for the first law of thermodynamics,

$$đQ = d\bar{E} + đW,$$

which converts the inexact differential $đQ$ into the exact differential dS. It follows that the entropy difference between any two macrostates and i can f be written

$$S_f - S_i = \int_i^f dS = \int_i^f \frac{đQ}{T},$$

where the integral is evaluated for any process through which the system is brought quasi-statically via a sequence of near-equilibrium configurations from its initial to its final macrostate. The process has to be quasi-static because the temperature T, which appears in the integrand, is only well-defined for an equilibrium state. Since the left-hand side of the above equation only depends on the initial and final states, it follows that the integral on the right-hand side is independent of the particular sequence of quasi-static changes used to get from to f. Thus, $\int_i^f đQ/T$ is independent of the process (provided that it is quasi-static).

All of the concepts which we have encountered up to now in this course, such as temperature, heat, energy, volume, pressure, etc., have been fairly familiar to us from other branches of Physics. However, entropy, which turns out to be of crucial importance in thermodynamics, is something quite new. Let us consider the following questions. What does the entropy of a system actually signify? What use is the concept of entropy?

ENTROPY

Consider an isolated system whose energy is known to lie in a narrow range. Let Ω be the number of accessible microstates. According to the principle of equal a priori probabilities, the system is equally likely to be found in any one of these states when it is in thermal equilibrium. The accessible states are just that set of microstates which are consistent with the macroscopic constraints imposed on the system.

These constraints can usually be quantified by specifying the values of some parameters $y_1, \ldots, y_n$ which characterize the macrostate. Note that these parameters are not necessarily external: e.g., we could specify either the volume (an external parameter) or the mean pressure (the mean force conjugate to the volume).

The number of accessible states is clearly a function of the chosen parameters, so we can write $\Omega \equiv \Omega(y_1, \ldots, y_n)$for the number of microstates consistent with a macrostate in which the general parameter y_α lies in the range y_α to $y_\alpha + dy_\alpha$.

Suppose that we start from a system in thermal equilibrium. According to statistical mechanics, each of the Ω_i, say, accessible states are equally likely. Let us now remove, or relax, some of the constraints imposed on the system. Clearly, all of the microstates formally accessible to the system are still accessible, but many additional states will, in general, become accessible. Thus, removing or relaxing constraints can only have the effect of increasing, or possibly leaving unchanged, the number of microstates accessible to the system. If the final number of accessible states is Ω_f, then we can write

$$\Omega_f \geq \Omega_i.$$

Immediately after the constraints are relaxed, the systems in the ensemble are not in any of the microstates from which they were previously excluded. So the systems only occupy a fraction

$$P_i = \frac{\Omega_i}{\Omega_f}$$

of the Ω_f states now accessible to them. This is clearly not a equilibrium situation. Indeed, if $\Omega_f \gg \Omega_i$ then the configuration in which the systems are only distributed over the original Ω_i states is an extremely unlikely one. In fact, its probability of occurrence is given by Eq. According to the theorem, *H* the ensemble will evolve in time until a more probable final state is reached in which the systems are evenly distributed over the Ω_f available states.

As a simple example, consider a system consisting of a box divided into two regions of equal volume. Suppose that, initially, one region is filled with gas and the other is empty. The constraint imposed on the system is, thus, that the coordinates of all of the molecules must lie within the filled region.

In other words, the volume accessible to the system $V = V_i$ is , where V_i is half the volume of the box. The constraints imposed on the system can be relaxed by removing the partition and allowing gas to flow into both regions. The volume accessible to the gas is now $V = V_f = 2V_i$. Immediately after the partition is removed, the system is in an extremely improbable state that at constant energy the variation of the number of accessible states of an ideal gas with the volume is

$$\Omega \propto V^N,$$

where N is the number of particles. Thus, the probability of observing the state immediately after the partition is removed in an ensemble of equilibrium systems with volume $V = V_f$ is

$$P_i = \frac{\Omega_i}{\Omega_f} = \left(\frac{V_i}{V_f}\right)^N = \left(\frac{1}{2}\right)^N.$$

If the box contains of order 1 mole of molecules then $N \sim 10^{24}$ and this probability is fantastically small:

$$P_i \sim \exp(-10^{24}).$$

Clearly, the system will evolve towards a more probable state.

This can also be phrased in terms of the parameters $y_1,...,y_n$ of the system. Suppose that a constraint is removed. For instance, one of the parameters, y, say, which originally had the value $y = y_i$, is now allowed to vary. According to statistical mechanics, all states accessible to the system are equally likely. So, the probability $P(y)$of finding the system in equilibrium with the parameter in the range to is just proportional y to $y + \delta y$ the number of microstates in this interval: i.e.,

$$P(y) \propto \Omega\,(y).$$

Usually, $\Omega(y)$ has a very pronounced maximum at some particular value $\tilde{y}$. This means that practically all systems in the final equilibrium ensemble have values of y close to $\tilde{y}$. Thus, if $y_i \neq \tilde{y}$ initially then the parameter y will change until it attains a final value close to $\tilde{y}$, where Ω is maximum.

If some of the constraints of an isolated system are removed then the parameters of the system tend to readjust themselves in such a way that

$$\Omega(y_1,...,y_n) \to \text{maximum}.$$

Suppose that the final equilibrium state has been reached, so that the systems in the ensemble are uniformly distributed over the Ω_f accessible final states. If the original constraints are reimposed then the systems in the ensemble still occupy these Ω_f states with equal probability. Thus, if $\Omega_f > \Omega_i$, simply restoring the constraints does not restore the initial situation. Once the systems are randomly distributed over the Ω_f states they cannot be expected to spontaneously move out of some of these states and occupy a more restricted class of states merely in response to the reimposition of a constraint. The initial condition can also not be restored by removing further constraints. This could only lead to even more states becoming accessible to the system.

Suppose that some process occurs in which an isolated system goes from some initial configuration to some final configuration. If the final configuration is such that the imposition or removal of constraints cannot by itself restore the initial condition then the process is deemed irreversible. On the other hand, if it is such that the imposition or removal of constraints can restore the initial condition then the process is deemed reversible.

From what we have already said, an irreversible process is clearly one in which the removal of constraints leads to a situation where $\Omega_f > \Omega_i$. A reversible process corresponds to the special case where the removal of constraints does not change the number of accessible states, so that $\Omega_f = \Omega_i$. In this situation, the systems remain distributed with equal probability over these states irrespective of whether the constraints are imposed or not.

Our microscopic definition of irreversibility is in accordance with the macroscopic . Recall that on a macroscopic level an irreversible process is one which "looks unphysical" when viewed in reverse. On a microscopic level it is clearly plausible that a system should spontaneously evolve from an improbable to a probable configuration in response to the relaxation of some constraint.

However, it is quite clearly implausible that a system should ever spontaneously evolve from a probable to an improbable configuration. Let us consider our example again. If a gas is initially restricted to one half of a box, via a partition, then the flow of gas from one side of the box to the other when the partition is removed is an irreversible process.

This process is irreversible on a microscopic level because the initial configuration cannot be recovered by simply replacing the partition. It is irreversible on a macroscopic level because it is obviously unphysical for the molecules of a gas to spontaneously distribute themselves in such a manner that they only occupy half of the available volume.

It is actually possible to quantify irreversibility. In other words, in addition to stating that a given process is irreversible, we can also give some indication of how irreversible it is. The parameter which measures irreversibility is just the number of accessible states Ω. Thus, if Ω for an isolated system spontaneously increases then the process is irreversible, the degree of irreversibility being proportional to the amount of the increase. If Ω stays the same then the process is reversible. Of course, it is unphysical for Ω to ever spontaneously decrease. In symbols, we can write

$$\Omega_f - \Omega_i \equiv \Delta\Omega \geq 0,$$

for any physical process operating on an isolated system. In practice, Ω itself is a rather unwieldy parameter with which to measure irreversibility. For instance, in the previous example, where an ideal gas doubles in volume (at constant energy) due to the removal of a partition, the fractional increase in Ω is

$$\frac{\Omega_f}{\Omega_i} \simeq 10^{2v\times 10^{23}},$$

where is the number of moles. This is an extremely large number! It is far more convenient to measure irreversibility in terms of ln Ω. If Eq. is true then it is certainly also true that

$$\ln \Omega_f - \ln \Omega_i \equiv \Delta \ln \Omega \geq 0$$

for any physical process operating on an isolated system. The increase in when an ideal gas doubles in volume (at constant energy) is

$$\ln \Omega_f - \ln \Omega_i = \nu N_A \ln 2,$$

where $N_A = 6 \times 10^{23}$. This is a far more manageable number! Since we usually deal with particles by the mole in laboratory physics, it makes sense to pre-multiply our measure of irreversibility by a number of order $1/N_A$. For historical reasons, the number which is generally used for this purpose is the Boltzmann constant k, which can be written

$$k = \frac{R}{N_A} \text{ joules/kelvin,}$$

where

$$R = 8.3143 \text{ joules/kelvin/mole}$$

is the ideal gas constant which appears in the well-known equation of state for an ideal gas, $PV = \upsilon RT$. Thus, the final form for our measure of irreversibility is

$$S = k \ln \Omega$$

This quantity is termed "entropy", and is measured in joules per degree kelvin. The increase in entropy when an ideal gas doubles in volume (at constant energy) is

$$S_f - S_i = \upsilon R \ln 2,$$

which is order unity for laboratory scale systems (i.e., those containing about one mole of particles). The essential irreversibility of macroscopic phenomena can be summed up as follows:

$$S_f - S_i = \Delta S \geq 0,$$

for a process acting on an isolated system. Thus, the entropy of an isolated system tends to increase with time and can never decrease. This proposition is known as the second law of thermodynamics.

One way of thinking of the number of accessible states Ω is that it is a measure of the disorder associated with a macrostate. For a system exhibiting a high degree of order we would expect a strong correlation between the motions of the individual particles. For instance, in a fluid there might be a strong tendency for the particles to move in one particular direction, giving rise to an ordered flow of the system in that direction. On the other hand, for a system exhibiting a low degree of order we expect far less correlation between the motions of individual particles.

It follows that, all other things being equal, an ordered system is more constrained than a disordered system, since the former is excluded from microstates in which there is not a strong correlation between individual particle motions, whereas the latter is not. Another way of saying this is that an ordered system has less accessible microstates than a corresponding disordered system. Thus, entropy is effectively a measure of the disorder in a system (the disorder increases with S). With this interpretation, the second law of thermodynamics reduces to the statement that isolated systems tend to become more disordered with time, and can never become more ordered.

Note that the second law of thermodynamics only applies to isolated systems. The entropy of a non-isolated system can decrease. For instance, if a gas expands (at constant energy) to twice its initial volume after the removal of a partition, we can subsequently recompress the gas to its original volume. The energy of the gas will increase because of the work done on it during compression, but if we absorb some heat from the gas then we can restore it to its initial state. Clearly, in restoring the gas to its original state, we have restored its original entropy.

This appears to violate the second law of thermodynamics because the entropy should have increased in what is obviously an irreversible process (just try to make a gas spontaneously occupy half of its original volume!). However, if we consider a new system consisting of the gas plus the compression and heat absorption machinery, then it is still true that the entropy of this system (which is assumed to be isolated) must increase in time.

Thus, the entropy of the gas is only kept the same at the expense of increasing the entropy of the rest of the system, and the total entropy is increased. If we consider the system of everything in the Universe, which is certainly an isolated system since there is nothing outside it with which it could interact, then the second law of thermodynamics becomes:

The disorder of the Universe tends to increase with time and can never decrease. An irreversible process is clearly one which increases the disorder of the Universe, whereas a reversible process neither increases nor decreases disorder. This definition is in accordance with our previous definition of an irreversible process as one which "does not look right" when viewed backwards.

One easy way of viewing macroscopic events in reverse is to film them, and then play the film backwards through a projector. There is a famous passage in the novel "Slaughterhouse 5," by Kurt Vonnegut, in which the hero, Billy

Pilgrim, views a propaganda film of an American World War II bombing raid on a German city in reverse. This is what the film appeared to show:

"American planes, full of holes and wounded men and corpses took off backwards from an airfield in England. Over France, a few German fighter planes flew at them backwards, sucked bullets and shell fragments from some of the planes and crewmen. They did the same for wrecked American bombers on the ground, and those planes flew up backwards and joined the formation.

The formation flew backwards over a German city that was in flames. The bombers opened their bomb bay doors, exerted a miraculous magnetism which shrunk the fires, gathered them into cylindrical steel containers, and lifted the containers into the bellies of the planes. The containers were stored neatly in racks. The Germans had miraculous devices of their own, which were long steel tubes. They used them to suck more fragments from the crewmen and planes. But there were still a few wounded Americans, though, and some of the bombers were in bad repair. Over France, though, German fighters came up again, made everything and everybody as good as new."

Vonnegut's point, I suppose, is that the morality of actions is inverted when you view them in reverse.

What is there about this passage which strikes us as surreal and fantastic? What is there that immediately tells us that the events shown in the film could never happen in reality? It is not so much that the planes appear to fly backwards and the bombs appear to fall upwards. After all, given a little ingenuity and a sufficiently good pilot, it is probably possible to fly a plane backwards.

Likewise, if we were to throw a bomb up in the air with just the right velocity we could, in principle, fix it so that the velocity of the bomb matched that of a passing bomber when their paths intersected. Certainly, if you had never seen a plane before it would not be obvious which way around it was supposed to fly. However, certain events are depicted in the film, "miraculous" events in Vonnegut's words, which would immediately strike us as the wrong way around even if we had never seen them before. For instance, the film might show thousands of white hot bits of shrapnel approach each other from all directions at great velocity, compressing an explosive gas in the process, which slows them down such that when they meet they fit together exactly to form a metal cylinder enclosing the gases and moving upwards at great velocity. What strikes us as completely implausible about this event is the spontaneous transition from the disordered motion of the gases and metal fragments to the ordered upward motion of the bomb.

Properties of Entropy

Entropy, as we have defined it, has some dependence on the resolution δE to which the energy of macrostates is measured. Recall that $\Omega(E)$ is the number of accessible microstates with energy in the range E to $E + \delta E$. Suppose that we choose a new resolution $\delta^* E$ and define a new density of states $\Omega^* E$ which is the number of states with energy in the range E to $E + \delta^* E$. It can easily be seen that

$$\Omega^*(E) = \frac{\delta^* E}{\delta E}\,\Omega(E).$$

It follows that the new entropy $S^* = k \ln \Omega^*$ is related to the previous entropy $S^* = k \ln \Omega$ via

$$S^* = S + k \ln \frac{\delta^* E}{\delta E}.$$

Now, our usual estimate that $\Omega \sim E^f$ gives $S \sim kf$, where f is the number of degrees of freedom. It follows that even if $\delta^* E$ were to differ from δE by of order f (i.e., twenty four orders of magnitude), which is virtually inconceivable, the second term on the right-hand side of the above equation is still only of order $k \ln f$, which is utterly negligible compared to kf. It follows that

$$S^* = S$$

to an excellent approximation, so our definition of entropy is completely insensitive to the resolution to which we measure energy (or any other macroscopic parameter).

Note that, like the temperature, the entropy of a macrostate is only well-defined if the macrostate is in equilibrium. The crucial point is that it only makes sense to talk about the number of accessible states if the systems in the ensemble are given sufficient time to thoroughly explore all of the possible microstates consistent with the known macroscopic constraints. In other words, we can only be sure that a given microstate is inaccessible when the systems in the ensemble have had ample opportunity to move into it, and yet have not done so. Note that for an equilibrium state, the entropy is just as well-defined as more familiar quantities such as the temperature and the mean pressure.

Consider, again, two systems A and A'which are in thermal contact but can do no work on one another. Let E and E' be the energies of the two systems, and $\Omega'(E)$ and $\Omega'(E')$ the respective densities of states. Furthermore, let $E^{(0)}$ be the conserved energy of the system as a whole and the $\Omega^{(0)}$ corresponding density of states. We have from Eq. that

$$\Omega^{(0)}\,(E) = \Omega(E)\,\Omega'(E'),$$

where $E' = E^{(0)} - E$. In other words, the number of states accessible to the whole system is the product of the numbers of states accessible to each subsystem, since every microstate of A can be combined with every microstate of $A'$to form a distinct microstate of the whole system that in equilibrium the mean energy of A takes the value $\bar{E} = \tilde{E}$ for which $\Omega^{(0)}(E)$ is maximum, and the temperatures of A and A' are equal. The distribution of E around the mean value is of order $\Delta^*E = \tilde{E}/\sqrt{f}$, where f is the number of degrees of freedom. It follows that the total number of accessible microstates is approximately the number of states which lie within Δ^*E of $\tilde{E}$. Thus,

$$\Omega_{tot}^{(0)} \simeq \frac{\Omega^{(0)}(\tilde{E})}{\delta E}\Delta^* E.$$

The entropy of the whole system is given by

$$S^{(0)} = k\ln\Omega_{tot}^{(0)} = k\ln\Omega^{(0)}(\tilde{E}) + k\ln\frac{\Delta^* E}{\delta E}.$$

According to our usual estimate, $\Omega \sim E^f$, the first term on the right-hand side is of order kf whereas the second term is

of order $k \ln (\tilde{E}/\sqrt{f}\,\delta E)$. Any reasonable choice for the energy subdivision δE should be greater than $\tilde{E}/f$, otherwise there would be less than one microstate per subdivision. It follows that the second term is less than or of order $k \ln f$, which is utterly negligible compared to kf. Thus,
$S^{(0)} = k \ln \Omega^{(0)}(\tilde{E}) = k\ln [\Omega(\tilde{E})\Omega(\tilde{E}')] = k \ln \Omega(\tilde{E}) + k \ln \Omega'(\tilde{E}')$
to an excellent approximation, giving

$$S^{(0)} = S(\tilde{E}) + S'(\tilde{E}').$$

It can be seen that the probability distribution for $\Omega^{(0)}(E)$is so strongly peaked around its maximum value that, for the purpose of calculating the entropy, the total number of states is equal to the maximum number of states [i.e., $\Omega^{(0)}_{tot} \sim \Omega^{(0)}(\tilde{E})$]. One consequence of this is that the entropy has the simple additive property Eq. Thus, the total entropy of two thermally interacting systems in equilibrium is the sum of the entropies of each system in isolation.

Uses of Entropy

We have defined a new function called entropy, denoted S, which parameterizes the amount of disorder in a macroscopic system. The entropy of an equilibrium macrostate is related to the number of accessible microstates Ω via

$$S = k \ln \Omega.$$

On a macroscopic level, the increase in entropy due to a quasi-static change in which an infinitesimal amount of heat $đQ$ is absorbed by the system is given by

$$dS = \frac{đQ}{T},$$

where T is the absolute temperature of the system. The second law of thermodynamics states that the entropy of an isolated system can never spontaneously decrease. Let us now briefly examine some consequences of these results.

Consider two bodies, A and A', which are in thermal contact but can do no work on one another. We know what is supposed to happen here. Heat flows from the hotter to the colder of the two bodies until their temperatures are the same. Consider a quasi-static exchange of heat between the two bodies. According to the first law of thermodynamics, if an infinitesimal amount of heat $đQ$ is absorbed by A then infinitesimal heat $đQ' = -đQ$ is absorbed by A'. The increase in the entropy of system A is $dS = đQ/T$ and the corresponding increase in the entropy of A' is $dS = đQ/T'$. Here, T and T' are the temperatures of the two systems, respectively. Note that $đQ$ is assumed to the sufficiently small that the heat transfer does not substantially modify the temperatures of either system. The change in entropy of the whole system is

$$dS^{(0)} = dS + dS' = \left(\frac{1}{T} - \frac{1}{T'}\right)đQ,$$

This change must be positive or zero, according to the second law of thermodynamics, so $dS^{(0)} \geq 0$. It follows that $đQ$ is positive (i.e., heat flows from A' to A) when $T' > T$, and vice versa. The spontaneous flow of heat only ceases when $T = T'$. Thus, the direction of spontaneous heat flow is a consequence of the second law of thermodynamics.

Note that the spontaneous flow of heat between bodies at different temperatures is always an irreversible process which increases the entropy, or disorder, of the Universe.

Consider, now, the slightly more complicated situation in which the two systems can exchange heat and also do work on one another via a movable partition. Suppose that the total volume is invariant, so that

$$V^{(0)} = V + V' \text{ constant,}$$

where V and V' are the volumes of A and A', respectively. Consider a quasi-static change in which system A absorbs an infinitesimal amount of heat $đQ$ and its volume simultaneously increases by an infinitesimal amount dV. The infinitesimal amount of work done by system A is $đW = \bar{p}dV$, where $\bar{p}$ is the mean pressure of A. According to the first law of thermodynamics,

$$đQ = dE + đW = dE + \bar{p}dV,$$

where dE is the change in the internal energy of A. Since $dS = đQ/T$, the increase in entropy of system A is written

$$dS = \frac{dE + \bar{p}dV}{T}.$$

Likewise, the increase in entropy of system A' is given by

$$dS' = \frac{dE' + \overline{p}' dV'}{T'}.$$

According to Eq.

$$\frac{1}{T} = \left(\frac{\partial S}{\partial E}\right)_V,$$

$$\frac{\overline{p}}{T} = \left(\frac{\partial S}{\partial V}\right)_E,$$

where the subscripts are to remind us what is held constant in the partial derivatives. We can write a similar pair of equations for the system A'.

The overall system is assumed to be isolated, so conservation of energy gives $dE + dE' = 0$. Furthermore, Eq. implies that $dV + dV' = 0$. It follows that the total change in entropy is given by

$$dS^{(0)} = dS + dS' = \left(\frac{1}{T} - \frac{1}{T'}\right) dE + \left(\frac{\overline{p}}{T} - \frac{\overline{p}'}{T'}\right) dV.$$

The equilibrium state is the most probable state. According to statistical mechanics, this is equivalent to the state with the largest number of accessible microstates. Finally, Eq. implies that this is the maximum entropy state.

The system can never spontaneously leave a maximum entropy state, since this would imply a spontaneous reduction in entropy, which is forbidden by the second law of thermodynamics.

A maximum or minimum entropy state must satisfy $dS^{(0)} = 0$ for arbitrary small variations of the energy and external parameters. It follows from Eq. that

$$T = T'$$

$$\overline{p} = \overline{p}',$$

for such a state. This corresponds to a maximum entropy state (i.e., an equilibrium state) provided

$$\left(\frac{\partial^2 S}{\partial E^2}\right)_V < 0,$$

$$\left(\frac{\partial^2 S}{\partial V^2}\right)_E < 0,$$

with a similar pair of inequalities for system A'. The usual estimate $\Omega \propto E^f V^f$, giving $S = kf \ln E + kf \ln V + \ldots$, ensures that the above inequalities are satisfied in conventional macroscopic systems. In the maximum entropy state the systems A and A' have equal temperatures (i.e., they are in thermal equilibrium) and equal pressures (i.e., they are in mechanical equilibrium). The second law of thermodynamics implies that the two interacting systems will evolve towards this state, and will then remain in it indefinitely (if left undisturbed).

ENTROPY AND QUANTUM MECHANICS

The entropy of a system is defined in terms of the number Ω of accessible microstates consistent with an overall energy in the range E to $E + \delta E$via

$$S = k \ln \Omega.$$

We have already demonstrated that this definition is utterly insensitive to the resolution δE to which the macroscopic energy is measured. In classical mechanics, if a system possesses f degrees of freedom then phase-space is conventionally subdivided into cells of arbitrarily chosen volume $h_0{}^f$. The number of accessible microstates is equivalent to the number of these cells in the volume of phase-space consistent with an overall energy of the system lying in the range E to $E + \delta E$. Thus,

$$\Omega = \frac{1}{h_0{}^f} \int \cdots \int dq_1 \cdots dq_f dp_1 \cdots dp_f,$$

giving

$$S = k \ln\left(\int \cdots \int dq_1 \cdots dq_f dp_1 \cdots dp_f\right) - kf \ln h_0.$$

Thus, in classical mechanics the entropy is undetermined to an arbitrary additive constant which depends on the size of the cells in phase-space. In fact, S increases as the cell size decreases. The second law of thermodynamics is only concerned with changes in entropy, and is, therefore, unaffected by an additive constant.

Likewise, macroscopic thermodynamical quantities, such as the temperature and pressure, which can be expressed as partial derivatives of the entropy with respect to various macroscopic parameters are unaffected by such a constant. So, in classical mechanics the entropy is rather like a gravitational potential: it is undetermined to an additive constant, but this does not affect any physical laws.

The non-unique value of the entropy comes about because there is no limit to the precision to which the state of a classical system can be specified.

In other words, the cell size h_0 can be made arbitrarily small, which corresponds to specifying the particle coordinates and momenta to arbitrary accuracy. However, in quantum mechanics the uncertainty principle sets a definite limit to how accurately the particle coordinates and momenta can be specified. In general,

$$\delta q_i \, \delta p_i \geq h_i$$

where p_i is the momentum conjugate to the generalized coordinate q_i, and δq_i, $\delta p_i$are the uncertainties in these quantities, respectively. In fact, in quantum mechanics the number of accessible quantum states with the overall energy in the range E to $E + \delta E$ is completely determined.

This implies that, in reality, the entropy of a system has a unique and unambiguous value.

Quantum mechanics can often be "mocked up" in classical mechanics by setting the cell size in phase-space equal to Planck's constant, so that $h_0 = h$.

This automatically enforces the most restrictive form of the uncertainty principle, $\delta q_i \delta p_i = h$.

In many systems, the substitution $h_0 \rightarrow h$ in Eq. gives the same, unique value for as that obtained from a full quantum mechanical calculation.

Consider a simple quantum mechanical system consisting of N non-interacting spinless particles of mass m confined in a cubic box of dimension L. The energy levels of the th particle are given by

$$e_i = \frac{\hbar^2\pi^2}{2mL^2}\left(n_{il}^2 + n_{i2}^2 + n_{i3}^2\right),$$

where n_{i1}, n_{i2}, and n_{i3} are three (positive) quantum numbers. The overall energy of the system is the sum of the energies of the individual particles, so that for a general state τ

$$E_r = \sum_{i=1}^{N} e_i.$$

The overall state of the system is completely specified by $3N$ quantum numbers, so the number of degrees of freedom is $f = 3N$. The classical limit corresponds to the situation where all of the quantum numbers are much greater than unity. In this limit, the number of accessible states varies with energy according to our usual estimate $\Omega \propto E^f$.

The lowest possible energy state of the system, the so-called ground-state, corresponds to the situation where all quantum numbers take their lowest possible value, unity. Thus, the ground-state energy E_0 is given by

$$E_0 = \frac{f\hbar^2\pi^2}{2mL^2}.$$

There is only one accessible microstate at the ground-state energy (i.e., that where all quantum numbers are unity), so by our usual definition of entropy

$$S(E_0) = k\ln 1 = 0.$$

In other words, there is no disorder in the system when all the particles are in their ground-states.

Clearly, as the energy approaches the ground-state energy, the number of accessible states becomes far less than the usual classical estimate E^f. This is true for all quantum mechanical systems. In general, the number of microstates varies roughly like

$$\Omega(E) \sim 1 + C(E - E_0)^f,$$

where C is a positive constant. According to Eq. the temperature varies approximately like

$$T \sim \frac{E - E_0}{kf},$$

provided $\Omega \gg 1$. Thus, as the absolute temperature of a system approaches zero, the internal energy approaches a limiting value E_0(the quantum mechanical ground-state energy), and the entropy approaches the limiting value zero. This proposition is known as the third law of thermodynamics.

At low temperatures, great care must be taken to ensure that equilibrium thermodynamical arguments are applicable, since the rate of attaining equilibrium may be very slow. Another difficulty arises when dealing with a system in which the atoms possess nuclear spins.

Typically, when such a system is brought to a very low temperature the entropy associated with the degrees of freedom not involving nuclear spins becomes negligible. Nevertheless, the number of microstates Ω_s corresponding to the possible nuclear spin orientations may be very large. Indeed, it may be just as large as the number of states at room temperature.

The reason for this is that nuclear magnetic moments are extremely small, and, therefore, have extremely weak mutual interactions. Thus, it only takes a tiny amount of heat energy in the system to completely randomize the spin orientations. Typically, a temperature as small as 10^{-3} degrees kelvin above absolute zero is sufficient to randomize the spins.

Suppose that the system consists of N atoms of spin 1/2. Each spin can have two possible orientations. If there is enough residual heat energy in the system to randomize the spins then each orientation is equally likely. If follows that there are $\Omega_s = 2^N$ accessible spin states.

The entropy associated with these states is $S_0 = k \ln \Omega_s = v R \ln 2$. Below some critical temperature, T_0, the interaction between the nuclear spins becomes significant, and the system settles down in some unique quantum mechanical ground-state (e.g., with all spins aligned).

In this situation, $S \to 0$, in accordance with the third law of thermodynamics. However, for temperatures which are small, but not small enough to "freeze out" the nuclear spin degrees of freedom, the entropy approaches a limiting value S_0 which depends only on the kinds of atomic nuclei in the system.

This limiting value is independent of the spatial arrangement of the atoms, or the interactions between them. Thus, for most practical purposes the third law of thermodynamics can be written

$$\text{as } T \to 0_+,\ S \to S_0,$$

where 0_+denotes a temperature which is very close to absolute zero, but still much larger than . This modification of the third law is useful because it can be applied at temperatures which are not prohibitively low.

Chapter 4

Applications of Statistical Thermodynamics

BOLTZMANN DISTRIBUTIONS

We have gained some understanding of the macroscopic properties of the air around us. For instance, we know something about its internal energy and specific heat capacity. How can we obtain some information about the statistical properties of the molecules which make up air? Consider a specific molecule: it constantly collides with its immediate neighbour molecules, and occasionally bounces off the walls of the room.

These interactions "inform" it about the macroscopic state of the air, such as its temperature, pressure, and volume. The statistical distribution of the molecule over its own particular microstates must be consistent with this macrostate. In other words, if we have a large group of such molecules with similar statistical distributions, then they must be equivalent to air with the appropriate macroscopic properties. So, it ought to be possible to calculate the probability distribution of the molecule over its microstates from a knowledge of these macroscopic properties.

We can think of the interaction of a molecule with the air in a classroom as analogous to the interaction of a small system *A* in thermal contact with a heat reservoir *A*¢.

The air acts like a heat reservoir because its energy fluctuations due to any interactions with the molecule are far too small to affect any of its macroscopic parameters. Let us determine the probability P_r of finding system A in one particular microstate $r$of energy E_r when it is thermal equilibrium with the heat reservoir A'.

As usual, we assume fairly weak interaction between A and A', so that the energies of these two systems are additive. The energy of A is not known at this stage. In fact, only the total energy of the combined system $A^{(0)} = A + A'$ is known. Suppose that the total energy lies in the range $E^{(0)}$ to $E^{(0)} + dE$. The overall energy is constant in time, since $A^{(0)}$ is assumed to be an isolated system, so

$$E_r + E' = E^{(0)},$$

where E' denotes the energy of the reservoir A'. Let $\Omega'(E')$ be the number of microstates accessible to the reservoir when its energy lies in the range E' to $E' + \delta E$ Clearly, if system A has an energy E_r then the reservoir $A¢$ must have an energy close to $E' = E^{(0)} - E_r$. Hence, since A is in one definite state (i.e., state r), and the total number of states accessible to A' is $\Omega'(E^{(0)} - E_r)$ it follows that the total number of states accessible to the combined system is simply $\Omega'(E^{(0)} - E_r)$. The principle of equal a priori probabilities tells us the the probability of occurrence of a particular situation is proportional to the number of accessible microstates. Thus,

$$P_r = C' \, \Omega' \, (E^{(0)} - E_r),$$

where C' is a constant of proportionality which is independent of r. This constant can be determined by the normalization condition

$$\sum_r P_r = 1,$$

where the sum is over all possible states of system A, irrespective of their energy.

Let us now make use of the fact that system A is far smaller than system A'. It follows that $E_r << E^{(0)}$, so the slowly varying logarithm of P_r can be Taylor expanded about $E' = E^{(0)}$. Thus,

$$\ln \text{Pr} = \ln C' + \ln \Omega'(E^{(0)}) - \left[\frac{\partial \ln \Omega'}{\partial E'}\right]_0 E_r + \cdots,$$

Note that we must expand $\ln P_r$, rather than P_r itself, because the latter function varies so rapidly with energy that the radius of convergence of its Taylor series is far too small for the series to be of any practical use. The higher order terms in Eq. can be safely neglected, because $E_r \ll E^{(0)}$. Now the derivative

$$\left[\frac{\partial \ln \Omega'}{\partial E'}\right]_0 \equiv \beta$$

is evaluated at the fixed energy $E' = E^{(0)}$, and is, thus, a constant independent of the energy E_r of A. In fact, we know that this derivative is just the temperature parameter $\beta = (kT)^{-1}$ characterizing the heat reservoir A'. Hence, Eq. becomes

$$\ln P_r = \ln C' + \ln \Omega'(E^{(0)}) - \beta E_r,$$

giving

$$P_r = C \exp(-\beta E_r),$$

where C is a constant independent of τ. The parameter C is determined by the normalization condition, which gives

$$C^{-1} = \sum_r \exp(-\beta E_r),$$

so that the distribution becomes

$$P_r = \frac{\exp(-\beta E_r)}{\sum_r \exp(-\beta E_r)}.$$

This is known as the Boltzmann probability distribution, and is undoubtably the most famous result in statistical physics.

The Boltzmann distribution often causes confusion. People who are used to the principle of equal a priori probabilities, which says that all microstates are equally probable, are understandably surprised when they come across the Boltzmann distribution which says that high energy microstates are markedly less probable then low energy states. However, there is no need for any confusion.

The principle of equal a priori probabilities applies to the whole system, whereas the Boltzmann distribution only applies to a small part of the system. The two results are perfectly consistent. If the small system is in a microstate with a comparatively high energy E_r then the rest of the system (i.e., the reservoir) has a slightly lower energy E' than usual (since the overall energy is fixed). The number of accessible microstates of the reservoir is a very strongly increasing function of its energy.

It follows that when the small system has a high energy then significantly less states than usual are accessible to the reservoir, and so the number of microstates accessible to the overall system is reduced, and, hence, the configuration is comparatively unlikely.

The strong increase in the number of accessible microstates of the reservoir with increasing E' gives rise to the strong (i.e., exponential) decrease in the likelihood of a state τ of the small system with increasing E_r. The exponential factor $\exp(-\beta\, E_r)$ is called the Boltzmann factor.

The Boltzmann distribution gives the probability of finding the small system A in one particular state τ of energy E_r. The probability $P(E)$ that A has an energy in the small range between E and $E + \delta E$ is just the sum of all the probabilities of the states which lie in this range. However, since each of these states has approximately the same Boltzmann factor this sum can be written

$$P(E) = C\Omega\,(E)\exp\,(-\beta E),$$

where $\Omega(E)$is the number of microstates of A whose energies lie in the appropriate range. Suppose that system $A$is itself a large system, but still very much smaller than system A'. For a large system, we expect $\Omega(E)$ to be a very rapidly increasing function of energy, so the probability $P(E)$ is the product of a rapidly increasing function of E and another rapidly decreasing function (i.e., the Boltzmann factor).

This gives a sharp maximum of $P(E)$ at some particular value of the energy. The larger system A, the sharper this maximum becomes. Eventually, the maximum becomes so sharp that the energy of system A is almost bound to lie at the most probable energy. As usual, the most probable energy is evaluated by looking for the maximum of $\ln P$, so

$$\frac{\partial \ln P}{\partial E} = \frac{\partial \ln \Omega}{\partial E} - \beta = 0,$$

giving

$$\frac{\partial \ln \Omega}{\partial E} = \beta.$$

Of course, this corresponds to the situation in which the temperature of A is the same as that of the reservoir. This is a result which we have seen before. Note, however, that the Boltzmann distribution is applicable no matter how small system A is, so it is a far more general result than any we have previously obtained.

PARAMAGNETISM

The simplest microscopic system which we can analyze using the Boltzmann distribution is one which has only two possible states (there would clearly be little point in analyzing a system with only one possible state). Most elements, and some compounds, are paramagnetic: i.e., their constituent atoms, or molecules, possess a permanent magnetic moment due to the presence of one or more unpaired electrons. Consider a substance whose constituent particles contain only one unpaired electron.

Such particles have spin 1/2, and consequently possess an intrinsic magnetic moment μ. According to quantum mechanics, the magnetic moment of a spin 1/2 particle can point either parallel or antiparallel to an external magnetic field B. Let us determine the mean magnetic moment $\bar{\mu}_B$ (in the direction of B) of the constituent particles of the substance when its absolute temperature is T. We assume, for the sake of simplicity, that each atom (or molecule) only interacts weakly with its neighbouring atoms. This enables us to focus attention on a single atom, and treat the remaining atoms as a heat bath at temperature T.

Our atom can be in one of two possible states: the (+) state in which its spin points up (i.e., parallel to B), and the (–) state in which its spin points down (i.e., antiparallel to B). In the (+)state, the atomic magnetic moment is parallel to the magnetic field, so that $\mu_\beta = \mu$. The magnetic energy of the atom is $\epsilon_+ = -\mu B$. In the (–) state, the atomic magnetic moment is antiparallel to the magnetic field, so that $\mu_B = -\mu$. The magnetic energy of the atom is $\in - = \mu_B$.

According to the Boltzmann distribution, the probability of finding the atom in the (+) state is

$$P_+ = C\exp(-\beta \in_+) = C\exp(\beta\,\mu\,\beta),$$

where C is a constant, and $\beta = (kT)^{-1}$. Likewise, the probability of finding the atom in the (–) state is

$$P_- = C\exp(-\beta \in_-) = C\exp(-\beta\,\mu\,B).$$

Clearly, the most probable state is the state with the lowest energy [i.e., the (+) state]. Thus, the mean magnetic moment points in the direction of the magnetic field (i.e., the atom is more likely to point parallel to the field than antiparallel).

It is clear that the critical parameter in a paramagnetic system is

$$y = \beta\mu B = \frac{\mu B}{kT}.$$

This parameter measures the ratio of the typical magnetic energy of the atom to its typical thermal energy. If the thermal energy greatly exceeds the magnetic energy then $y \ll 1$,, and the probability that the atomic moment points parallel to the magnetic field is about the same as the probability that it points antiparallel. In this situation, we expect the mean atomic moment to be small, so that $\bar{\mu}_0 \simeq 0$. On the other hand, if the magnetic energy greatly exceeds the thermal energy then $y \gg 1$, and the atomic moment is far more likely to point parallel to the magnetic field than antiparallel. In this situation, we expect $\bar{\mu}_n \simeq \mu$.

Let us calculate the mean atomic moment $\bar{\mu}_n$. The usual definition of a mean value gives

$$\bar{\mu}_B = \frac{P_+\mu + P_-(-\mu)}{P_+ + P_-} = \mu\frac{\exp(\beta\mu\beta) - \exp(-\beta\mu\beta)}{\exp(\beta\mu\beta) + \exp(-\beta\mu\beta)}.$$

This can also be written

$$\bar{\mu}_B = \mu \tan h\,\frac{\mu B}{kT},$$

where the hyperbolic tangent is defined

$$\tanh y \frac{\exp(y)-\exp(-y)}{\exp(y)+\exp(-y)}.$$

For small arguments $y \ll 1$,

$$\tanh y \simeq y - \frac{y^3}{3} + \cdots,$$

whereas for large arguments $y \gg 1$,

$$\tanh \simeq 1.$$

It follows that at comparatively high temperatures, $kT \gg \mu B$,

$$\bar{\mu}_B \simeq \frac{\mu^2 B}{kT},$$

whereas at comparatively low temperatures, $kT \ll \mu B$,

$$\bar{\mu}_B \simeq \mu.$$

Suppose that the substance contains N_0 atoms (or molecules) per unit volume. The magnetization is defined as the mean magnetic moment per unit volume, and is given by

$$\bar{M}_0 = N_0\, \bar{\mu}_B.$$

At high temperatures, $kT \gg \mu B$, the mean magnetic moment, and, hence, the magnetization, is proportional to the applied magnetic field, so we can write

$$\bar{M}_0 \simeq \chi B,$$

where χ is a constant of proportionality known as the magnetic susceptibility. It is clear that the magnetic susceptibility of a spin 1/2 paramagnetic substance takes the form

$$\chi = \frac{N_0 \mu^2}{kT}.$$

The fact that $\chi \propto T^{-1}$ is known as Curie's law, because it was discovered experimentally by Pierre Curie at the end of the nineteenth century. At low temperatures, $kT \ll \mu B$,

$$\bar{M}_0 \to N_0 \mu,$$

so the magnetization becomes independent of the applied field.

This corresponds to the maximum possible magnetization, where all atomic moments are lined up parallel to the field. The breakdown of the $\bar{M}_0 \propto B$ law at low temperatures (or high magnetic fields) is known as saturation.

The above analysis is only valid for paramagnetic substances made up of spin one-half (J = 1/2) atoms or molecules. However, the analysis can easily be generalized to take account of substances whose constituent particles possess higher spin (i.e., J > 1/2).

Compares the experimental and theoretical magnetization versus field-strength curves for three different substances made up of spin 3/2, spin 5/2, and spin 7/2 particles, showing the excellent agreement between the two sets of curves. Note that, in all cases, the magnetization is proportional to the magnetic field-strength at small field-strengths, but saturates at some constant value as the field-strength increases.

The previous analysis completely neglects any interaction between the spins of neighbouring atoms or molecules. It turns out that this is a fairly good approximation for paramagnetic substances.

However, for ferromagnetic substances, in which the spins of neighbouring atoms interact very strongly, this approximation breaks down completely. Thus, the above analysis does not apply to ferromagnetic substances.

Mean Values

Consider a system in contact with a heat reservoir. The systems in the representative ensemble are distributed over their accessible states in accordance with the Boltzmann distribution. Thus, the probability of occurrence of some state τ with energy E_r is given by

$$P_r = \frac{\exp(-\beta\, E_r)}{\sum_r \exp(-\beta E_r)}.$$

The mean energy is written

$$\bar{E} = \frac{\sum_r \exp(-\beta E_r)\, E_r}{\sum_r \exp(-\beta E_r)},$$

where the sum is taken over all states of the system, irrespective of their energy. Note that

$$\sum_r \exp(-\beta E_r)E_r = -\sum_r \frac{\partial}{\partial\beta}\exp(-\beta E_r) = -\frac{\partial Z}{\partial\beta},$$

where

$$Z - \sum_r \exp(-\beta E_r).$$

It follows that

$$\overline{E} = -\frac{1}{Z}\frac{\partial Z}{\partial\beta} = -\frac{\partial \ln Z}{\partial\beta}.$$

The quantity Z, which is defined as the sum of the Boltzmann factor over all states, irrespective of their energy, is called the partition function. We have just demonstrated that it is fairly easy to work out the mean energy of a system using its partition function. In fact, as we shall discover, it is easy to calculate virtually any piece of statistical information using the partition function.

Let us evaluate the variance of the energy. We know that

$$\overline{(\Delta E)^2} = \overline{E^2} - \overline{E}^2.$$

Now, according to the Boltzmann distribution,

$$\overline{E^2} = \frac{\sum_r \exp(-\beta E_r)\, E_r{}^2}{\sum_r \exp(-\beta E_r)}.$$

However,

$$\sum_r \exp(-\beta E_r)E_r{}^2 = -\frac{\partial}{\partial\beta}\left[\sum_r \exp(-E_r)E_r\right] = \left(-\frac{\partial}{\partial\beta}\right)^2\left[\sum_r \exp(-\beta E_r)\right].$$

Hence,

$$\overline{E^2} = \frac{1}{Z}\frac{\partial^2 Z}{\partial\beta^2}.$$

We can also write

$$\overline{E^2} = \frac{\partial}{\partial\beta}\left(\frac{1}{Z}\frac{\partial Z}{\partial\beta}\right) + \frac{1}{Z^2}\left(\frac{\partial Z}{\partial\beta}\right)^2 = -\frac{\partial\overline{E}}{\partial\beta} + \overline{E}^2,$$

where use has been made of Eq. It follows from Eq. that

$$\overline{(\Delta E)^2} = -\frac{\partial \bar{E}}{\partial \beta} = \frac{\partial^2 \ln Z}{\partial \beta^2}.$$

Thus, the variance of the energy can be worked out from the partition function almost as easily as the mean energy. Since, by definition, a variance can never be negative, it follows that $\partial \bar{E} / \partial \beta \leq 0$, or, equivalently, $\partial \bar{E} / \partial T \geq 0$. Hence, the mean energy of a system governed by the Boltzmann distribution always increases with temperature.

Suppose that the system is characterized by a single external parameter x (such as its volume). The generalization to the case where there are several external parameters is obvious. Consider a quasi-static change of the external parameter from x to $x + dx$. In this process, the energy of the system in state r changes by

$$\delta E_r = -\frac{\partial E_r}{\partial x} dx.$$

The macroscopic work $đW$ done by the system due to this parameter change is

$$đW = \frac{\sum_r \exp(-E_r)(-\partial E_r / \partial x \, dx)}{\sum_r \exp(-\beta E_r)}.$$

In other words, the work done is minus the average change in internal energy of the system, where the average is calculated using the Boltzmann distribution. We can write

$$\sum_r \exp(-\beta E_r) \frac{\partial E_r}{\partial x} = -\frac{1}{\beta} \frac{\partial}{\partial x} \left[\sum_r \exp(-\beta E_r) \right] = -\frac{1}{\beta} \frac{\partial Z}{\partial x},$$

which gives

$$đW = \frac{1}{\beta Z} \frac{\partial Z}{\partial x} dx = \frac{1}{\beta} \frac{\partial \ln Z}{\partial x} dx.$$

We also have the following general expression for the work done by the system

$$đW = \overline{X} dx,$$

where

$$\overline{X} = -\overline{\frac{\partial E_r}{\partial x}}$$

is the mean generalized force conjugate to x. It follows that

$$\overline{X} = -\frac{1}{\beta}\frac{\partial \ln Z}{\partial x}.$$

Suppose that the external parameter is the volume, so $x = V$. It follows that

$$đW = \overline{p}dV = \frac{1}{\beta}\frac{\partial \ln Z}{\partial V}dV.$$

and

$$\overline{p} = \frac{1}{\beta}\frac{\partial \ln Z}{\partial V}.$$

Since the partition function is a function of β and V(the energies E_r depend on V), it is clear that the above equation relates the mean pressure $\overline{p}$ to T (via $\beta = 1/kT$) and V. In other words, the above expression is the equation of state. Hence, we can work out the pressure, and even the equation of state, using the partition function.

Partition Functions

It is clear that all important macroscopic quantities associated with a system can be expressed in terms of its partition function Z. Let us investigate how the partition function is related to thermodynamical quantities. Recall that Z is a function of both β and x (where x is the single external parameter). Hence, $Z = Z(\beta, x)$, and we can write

$$d\ln Z = \frac{\partial \ln Z}{\partial x}dx + \frac{\partial \ln Z}{\partial x}d\beta.$$

Consider a quasi-static change by which x and βchange so slowly that the system stays close to equilibrium, and, thus, remains distributed according to the Boltzmann distribution. If follows from Eqs. that

$$d\ln Z = \beta\, đW - \overline{E}\, d\beta.$$

The last term can be rewritten

$$d\ln Z = \beta đW - d(\bar{E}\,\beta) + \beta\, d\bar{E},$$

giving

$$d(\ln Z + \beta\bar{E}) = \beta(đW + d\bar{E}) \equiv \beta đQ.$$

The above equation shows that although the heat absorbed by the system $đQ$ is not an exact differential, it becomes one when multiplied by the temperature parameter β. This is essentially the second law of thermodynamics. In fact, we know that

$$dS = \frac{đQ}{T}.$$

Hence,

$$S \equiv k(\ln Z + \beta\bar{E}).$$

This expression enables us to calculate the entropy of a system from its partition function.

Suppose that we are dealing with a system $A^{(0)}$ consisting of two systems A and A'which only interact weakly with one another. Let each state of A be denoted by an index $\tau$and have a corresponding energy E_r. Likewise, let each state of A' be denoted by an index s and have a corresponding energy $E's$. A state of the combined system $A^{(0)}$ is then denoted by two indices τ and s. Since A and A' only interact weakly their energies are additive, and the energy of state τs is

$$E_{\tau s}^{(0)} = E_r + E_s^{'}.$$

By definition, the partition function of $A^{(0)}$ takes the form

$$Z^{(0)} = \sum_{\tau,s} \exp[-\beta\, E_{\tau s}^{(0)}]$$

$$= \sum_{\tau,s} \exp(-\beta\,[E_r + E_s^{'}])$$

$$= \left[\sum_{\tau} \exp(-\beta\, E_r)\right]\left[\sum_{s} \exp(-\beta\, E_s^{'})\right].$$

Hence,

$$Z^{(0)} = ZZ',$$

giving

$$\ln Z^{(0)} = \ln Z + \ln Z',$$

where Z and Z' are the partition functions of A and A', respectively. It follows from Eq. that the mean energies of $A^{(0)}$, A, and A' are related by

$$\overline{E}^{(0)} = \overline{E} + \overline{E}'.$$

It also follows from Eq. that the respective entropies of these systems are related via

$$S^{(0)} = S + S'.$$

Hence, the partition function tells us that the extensive thermodynamic functions of two weakly interacting systems are simply additive.

It is clear that we can perform statistical thermodynamical calculations using the partition function Zinstead of the more direct approach in which we use the density of states Ω. The former approach is advantageous because the partition function is an unrestricted sum of Boltzmann factors over all accessible states, irrespective of their energy, whereas the density of states is a restricted sum over all states whose energies lie in some narrow range.

In general, it is far easier to perform an unrestricted sum than a restricted sum.

Thus, it is generally easier to derive statistical thermodynamical results using Zrather than Ω, although Ω has a far more direct physical significance than Z.

IDEAL MONATOMIC GASES

Let us now practice calculating thermodynamic relations using the partition function by considering an example with which we are already quite familiar: i.e., an ideal monatomic gas. Consider a gas consisting of *N*identical monatomic molecules of mass *m* enclosed in a container of volume *V*. Let us denote the position and momentum vectors of the *i*th molecule by r_i and p_i, respectively.

Since the gas is ideal, there are no interatomic forces, and the total energy is simply the sum of the individual kinetic energies of the molecules:

$$E = \sum_{i=1}^{N} \frac{p_i^{\,2}}{2m},$$

where $p_i^2 = p_i \cdot p_i$.

Let us treat the problem classically. In this approach, we divide up phase-space into cells of equal volume h_0^f. Here, f is the number of degrees of freedom, and h_0 is a small constant with dimensions of angular momentum which parameterizes the precision to which the positions and momenta of molecules are determined. Each cell in phase-space corresponds to a different state. The partition function is the sum of the Boltzmann factor exp $(-\beta\, E_r)$ over all possible states, where E_r is the energy of state τ. Classically, we can approximate the summation over cells in phase-space as an integration over all phase-space. Thus,

$$Z = \int \cdots \int \exp(-\beta\, E) \frac{d^3 r_1 \cdots d^3 r_N\, d^3 p_1 \cdots d^3{}_{PN}}{h_0^{\,3N}},$$

where $3N$ is the number of degrees of freedom of a monatomic gas containing N molecules. Making use of Eq. the above expression reduces to

$$Z = \frac{V^N}{h_0^{\,3N}} \int \cdots \int \exp[-\beta/2m) p_1^{\,2}]\, d^3 p_1 \cdots \exp[-(\beta/2m) p_N^2] d^3{}_{PN}.$$

Note that the integral over the coordinates of a given molecule simply yields the volume of the container, V, since the energy E is independent of the locations of the molecules in an ideal gas. There are N such integrals, so we obtain the factor V^N in the above expression. Note, also, that each of the integrals over the molecular momenta in Eq. are identical: they differ only by irrelevant dummy variables of integration. It follows that the partition function Z of the gas is made up of the product of N identical factors: i.e.,

$$Z = \zeta^N,$$

where

$$\zeta = \frac{V}{h_0^3} \int \exp[-(\beta/2m) p^2] d^3 p$$

is the partition function for a single molecule. Of course, this result is obvious, since we have already shown that the partition function for a system made up of a number of weakly interacting subsystems is just the product of the partition functions of the subsystems.

The integral in Eq. is easily evaluated:

$$\int \exp[-(\beta/2m)p^2]d^3p = \int_{-\infty}^{\infty} \exp[-(\beta/2m)\,p_x{}^2]dp_x$$

$$\int_{-\infty}^{\infty} \exp[-(\beta/2m)\,p_y{}^2]dp_y \times \int_{-\infty}^{\infty} \exp[-(\beta/2m)\,p_z{}^2]dp_z$$

$$= \left(\sqrt{\frac{2\pi m}{\beta}}\right)^3,$$

where use has been made of Eq. Thus,

$$\zeta = V\left(\frac{2\pi m}{h_0{}^2\beta}\right)^{3/2},$$

and

$$\ln Z = N\ln\zeta = N\left[\ln V - \frac{3}{2}\ln\beta + \frac{3}{2}\ln\left(\frac{2\pi m}{h_0{}^2}\right)\right].$$

The expression for the mean pressure yields

$$\overline{p} = \frac{1}{\beta}\frac{\partial \ln Z}{\partial V} = \frac{1}{\beta}\frac{N}{V},$$

which reduces to the ideal gas equation of state

$$pV = NkT = vRT,$$

where use has been made of $N = vN_A$ and $R = N_A k$. According to Eq. the mean energy of the gas is given by

$$\overline{E} = -\frac{\partial \ln Z}{\partial\beta} = \frac{3}{2}\frac{N}{\beta} = v\frac{3}{2}RT.$$

Note that the internal energy is a function of temperature alone, with no dependence on volume. The molar heat capacity at constant volume of the gas is given by

$$c_v = \frac{1}{v}\left(\frac{\partial \overline{E}}{\partial T}\right)_V = \frac{3}{2}R,$$

so the mean energy can be written

$$\bar{E} = v\, c_v T.$$

We have seen all of the above results before. Let us now use the partition function to calculate a new result. The entropy of the gas can be calculated quite simply from the expression

$$S = k(\ln Z + \beta \bar{E}).$$

Thus,

$$S = vR\left[\ln V - \frac{3}{2}\ln\beta + \frac{3}{2}\ln\left(\frac{2\pi m}{h_0^{\,2}}\right) + \frac{3}{2}\right],$$

or

$$S = vR\left[\ln V + \frac{3}{2}\ln T + \sigma\right],$$

where

$$\sigma = \frac{3}{2}\ln\left(\frac{2\pi m k}{h_0^{\,2}}\right) + \frac{3}{2}.$$

The above expression for the entropy of an ideal gas is certainly new. Unfortunately, it is also quite obviously incorrect.

GIBB'S PARADOX

What has gone wrong? First of all, let us be clear why Eq. is incorrect.

We can see that $S \to -\infty$ as $T \to 0$, which contradicts the third law of thermodynamics. However, this is not a problem. Equation was derived using classical physics, which breaks down at low temperatures. Thus, we would not expect this equation to give a sensible answer close to the absolute zero of temperature.

Equation is wrong because it implies that the entropy does not behave properly as an extensive quantity. Thermodynamic quantities can be divided into two groups, extensive and intensive. Extensive quantities increase by a factor α when the size of the system under consideration is increased by the same factor. Intensive quantities stay the same. Energy and volume are typical extensive quantities. Pressure and temperature are typical intensive quantities. Entropy is very definitely an extensive quantity. We have shown that the entropies of two weakly interacting systems are additive. Thus, if we double the size of a system we expect the entropy to double as well. Suppose that we have a system of volume V containing v moles of ideal gas at temperature T. Doubling the size of the system is like joining two identical systems together to form a new system of volume $2V$ containing $2v$ moles of gas at temperature T. Let

$$S = vR\left[\ln V + \frac{3}{2}\ln T + \sigma\right]$$

denote the entropy of the original system, and let

$$S' = 2vR\left[\ln 2V + \frac{3}{2}\ln T + \sigma\right]$$

denote the entropy of the double-sized system. Clearly, if entropy is an extensive quantity (which it is!) then we should have

$$S' = 2S.$$

But, in fact, we find that

$$S' - 2S = 2vR\ln 2.$$

So, the entropy of the double-sized system is more than double the entropy of the original system.

Where does this extra entropy come from? Well, let us consider a little more carefully how we might go about doubling the size of our system. Suppose that we put another identical system adjacent to it, and separate the two systems by a partition.

Let us now suddenly remove the partition. If entropy is a properly extensive quantity then the entropy of the overall system should be the same before and after the partition is removed. It is certainly the case that the energy (another extensive quantity) of the overall system stays the same. However, according to Eq. the overall entropy of the system increases by $2vR$ ln 2 after the partition is removed. Suppose, now, that the second system is identical to the first system in all respects except that its molecules are in some way slightly different to the molecules in the first system, so that the two sets of molecules are distinguishable. In this case, we would certainly expect an overall increase in entropy when the partition is removed. Before the partition is removed, it separates type 1 molecules from type 2 molecules. After the partition is removed, molecules of both types become jumbled together.

This is clearly an irreversible process. We cannot imagine the molecules spontaneously sorting themselves out again. The increase in entropy associated with this jumbling is called entropy of mixing, and is easily calculated. We know that the number of accessible states of an ideal gas varies with volume like $\Omega \propto V^N$. The volume accessible to type 1 molecules clearly doubles after the partition is removed, as does the volume accessible to type 2 molecules. Using the fundamental formula $S = k \ln \Omega$, the increase in entropy due to mixing is given by

$$S = 2k \ln \frac{\Omega_f}{\Omega_i} = 2Nk \ln \frac{V_f}{V_i} = 2v\, R \ln 2.$$

It is clear that the additional entropy $2vR \ln 2$, which appears when we double the size of an ideal gas system by joining together two identical systems, is entropy of mixing of the molecules contained in the original systems. But, if the molecules in these two systems are indistinguishable, why should there be any entropy of mixing? Well, clearly, there is no entropy of mixing in this case.

At this point, we can begin to understand what has gone wrong in our calculation. We have calculated the partition function assuming that all of the molecules in our system have the same mass and temperature, but we have never explicitly taken into account the fact that we consider the molecules to be indistinguishable.

In other words, we have been treating the molecules in our ideal gas as if each carried a little license plate, or a social security number, so that we could always tell one from another. In quantum mechanics, which is what we really should be using to study microscopic phenomena, the essential indistinguishability of atoms and molecules is hard-wired into the theory at a very low level. Our problem is that we have been taking the classical approach a little too seriously. It is plainly silly to pretend that we can distinguish molecules in a statistical problem, where we do not closely follow the motions of individual particles. A paradox arises if we try to treat molecules as if they were distinguishable. This is called Gibb's paradox, after the American physicist Josiah Gibbs who first discussed it. The resolution of Gibb's paradox is quite simple: treat all molecules of the same species as if they were indistinguishable. In our previous calculation of the ideal gas partition function, we inadvertently treated each of the N molecules in the gas as distinguishable.

Because of this, we overcounted the number of states of the system. Since the $N!$ possible permutations of the molecules amongst themselves do not lead to physically different situations, and, therefore, cannot be counted as separate states, the number of actual states of the system is a factor $N!$ less than what we initially thought. We can easily correct our partition function by simply dividing by this factor, so that

$$Z = \frac{\zeta^N}{N!}.$$

This gives

$$\ln Z = N \ln \zeta - N!,$$

or

$$\ln Z = N \ln \zeta - N \ln N + N,$$

using Stirling's approximation. Note that our new version of ln Z differs from our previous version by an additive term involving the number of particles in the system. This explains why our calculations of the mean pressure and mean energy, which depend on partial derivatives of ln Z with respect to the volume and the temperature parameter β, respectively, came out all right. However, our expression for the entropy S is modified by this additive term. The new expression is

$$S = vR\left[\ln V - \frac{3}{2}\ln\beta + \frac{3}{2}\ln\left(\frac{2\pi\, mk}{h_0^{\,2}}\right) + \frac{3}{2}\right] + k(-N\ln N + N).$$

This gives

$$S = vR\left[\ln\frac{V}{N} + \frac{3}{2}\ln T + \sigma_0\right]$$

where

$$\sigma_0 = \frac{3}{2}\ln\left(\frac{2\pi\, mk}{h_0^{\,2}}\right) + \frac{5}{2}.$$

It is clear that the entropy behaves properly as an extensive quantity in the above expression: i.e., it is multiplied by a factor α when v, V, and N are multiplied by the same factor.

THE EQUIPARTITION THEOREM

The internal energy of a monatomic ideal gas containing N particles. This means that each particle possess, on average, units of energy. Monatomic particles have only three translational degrees of freedom, corresponding to their motion in three dimensions. They possess no internal rotational or vibrational degrees of freedom. Thus, the mean energy per degree of freedom in a monatomic ideal gas is. In fact, this is a special case of a rather general result. Let us now try to prove this.

Suppose that the energy of a system is determined by some generalized coordinates and corresponding generalized momenta Pk, so that

$$E = E(q_1, ..., q_f, p_1, ..., p_f).$$

Suppose further that:

The total energy splits additively into the form

$$E = \in_i (p_i) + E'(q_1, ..., p_f),$$

where $\in_i$ involves only one variable p_i, and the remaining part E' does not depend on p_i.

The function $\in_i$ is quadratic in p_i, so that

$$\in_i (p_i) + bp_i^2,$$

where b is a constant.

The most common situation in which the above assumptions are valid is where p_i is a momentum. This is because the kinetic energy is usually a quadratic function of each momentum component, whereas the potential energy does not involve the momenta at all. However, if a coordinate q_i were to satisfy assumptions 1 and 2 then the theorem we are about to establish would hold just as well.

What is the mean value of ϵ_i in thermal equilibrium if conditions 1 and 2 are satisfied? If the system is in equilibrium at absolute temperature $T \equiv (k\beta)^{-1}$then it is distributed according to the Boltzmann distribution. In the classical approximation, the mean value of ϵ_i is expressed in terms of integrals over all phase-space:

$$\bar{\epsilon}_i = \frac{\int_{-\infty}^{\infty} \exp[-\beta E(q_1,...,p_f)]\, \epsilon_i \, dq_1...dp_f}{\int_{-\infty}^{\infty} \exp[-\beta E(q_1,...,p_f)] dq_1...dp_f}.$$

Condition 1 gives

$$\bar{\epsilon}_i = \frac{\int_{-\infty}^{\infty} \exp[-\beta(\epsilon_i + E')]\, \epsilon_i \, dq_1...dp_f}{\int_{-\infty}^{\infty} \exp[-\beta(\epsilon_i + E')] dq_1...dp_f}$$

$$= \frac{\int_{-\infty}^{\infty} \exp(-\beta \epsilon_i)\, \epsilon_i \, dp_i \int_{-\infty}^{\infty} \exp(-\beta E')\, dq_1...dp_f}{\int_{-\infty}^{\infty} \exp(-\beta \epsilon_i)\, dp_i \int_{-\infty}^{\infty} \exp(-\beta E')\, dq_1...dp_f},$$

where use has been made of the multiplicative property of the exponential function, and where the last integrals in both the numerator and denominator extend over all variables *qk*and *pk* except p_i. These integrals are equal and, thus, cancel. Hence,

$$\bar{\epsilon}_i = \frac{\int_{-\infty}^{\infty} \exp(-\beta \epsilon_i)\, \epsilon_i \, dq_i}{\int_{-\infty}^{\infty} \exp(-\beta \epsilon_i) dp_i}.$$

This expression can be simplified further since

$$\int_{-\infty}^{\infty} \exp(-\beta \epsilon_i)\, \epsilon_i \, dp_i \equiv -\frac{\theta}{\partial\beta}\left[\int_{-\infty}^{\infty} \exp(-\beta \epsilon_i) dp_i\right],$$

so

$$\bar{\epsilon}_i = -\frac{\partial}{\partial \beta} \ln\left[\int_{\infty}^{\infty} \exp(-\beta \epsilon_i) dp_i\right].$$

According to condition 2,

where $y = \sqrt{\beta}\ p_i$. Thus,

$$\ln \int_{-\infty}^{\infty} \exp(-\beta \epsilon_i)\, dp_i = -\frac{1}{2} \ln \beta + \ln \int_{-\infty}^{\infty} \exp(-by^2)\, dy.$$

Note that the integral on the right-hand side does not depend on β at all. It follows from Eq. that

$$\bar{\epsilon}_i = -\frac{\partial}{\partial \beta}\left(-\frac{1}{2} \ln \beta\right) = \frac{1}{2\beta},$$

giving

$$\bar{\epsilon}_i = \frac{1}{2} kT.$$

This is the famous equipartition theorem of classical physics. It states that the mean value of every independent quadratic term in the energy is equal to $(1/2)kT$. If all terms in the energy are quadratic then the mean energy is spread equally over all degrees of freedom (hence the name "equipartition").

HARMONIC OSCILLATORS

Our proof of the equipartition theorem depends crucially on the classical approximation. To see how quantum effects modify this result, let us examine a particularly simple system which we know how to analyze using both classical and quantum physics: i.e., a simple harmonic oscillator. Consider a one-dimensional harmonic oscillator in equilibrium with a heat reservoir at temperature T. The energy of the oscillator is given by

$$E = \frac{p^2}{2m} + \frac{1}{2} kx^2,$$

where the first term on the right-hand side is the kinetic energy, involving the momentum p and mass m, and the second term is the potential energy, involving the displacement x and the force constant k. Each of these terms is quadratic in the respective variable. So, in the classical approximation the equipartition theorem yields:

$$\frac{\overline{p^2}}{2m} = \frac{1}{2}kT,$$

$$\frac{1}{2}k\overline{x^2} = \frac{1}{2}kT.$$

That is, the mean kinetic energy of the oscillator is equal to the mean potential energy which equals (1/2)kT. It follows that the mean total energy is

$$\overline{E} = \frac{1}{2}kT + \frac{1}{2}kT = kT.$$

According to quantum mechanics, the energy levels of a harmonic oscillator are equally spaced and satisfy

$$E_n = (n + 1/2)\,\hbar\omega,$$

where n is a non-negative integer, and

$$\omega = \sqrt{\frac{k}{m}}.$$

The partition function for such an oscillator is given by

$$Z = \sum_{n=0}^{\infty} \exp(-\beta E_n) = \exp[-(1/2)\beta\hbar\omega]\sum_{n=0}^{\infty} \exp(-n\beta\hbar\omega).$$

Now,

$$\sum_{n=0}^{\infty} \exp(-n\beta\hbar\omega) = 1 + \exp(-\beta\hbar\omega) + \exp(-2\,\beta\hbar\omega) + \cdots$$

is simply the sum of an infinite geometric series, and can be evaluated immediately,

$$\sum_{n=0}^{\infty} \exp(-n\beta\hbar\omega) = \frac{1}{1 - \exp(-\beta\hbar\omega)}.$$

Thus, the partition function takes the form

$$Z = \frac{\exp[-(1/2)\beta\hbar\omega]}{1-\exp(-\beta\hbar\omega)},$$

and

$$\ln Z = -\frac{1}{2}\beta\hbar\omega - \ln[1-\exp(-\beta\hbar\omega)].$$

The mean energy of the oscillator is given by Eq.

$$\bar{E} = -\frac{\partial}{\partial\beta}\ln Z = -\left[-\frac{1}{2}\hbar\omega - \frac{\exp(-\beta\hbar\omega)\,\hbar\omega}{1-\exp(-\beta\hbar\omega)}\right],$$

or

$$\bar{E} = \hbar\omega\left[-\frac{1}{2} + \frac{1}{\exp(\beta\hbar\omega)-1}\right].$$

Consider the limit

$$\beta\hbar\omega = \frac{\hbar\omega}{kT} \ll 1,$$

in which the thermal energy kT is large compared to the separation $\hbar\omega$ between the energy levels. In this limit,

$$\exp(\beta\hbar\omega) \simeq 1 + \beta\hbar\omega,$$

so

$$\bar{E} \simeq \hbar\omega\left[\frac{1}{2} + \frac{1}{\beta\hbar\omega}\right] \simeq \hbar\omega\left[\frac{1}{\beta\hbar\omega}\right],$$

giving

$$\bar{E} \simeq \frac{1}{\beta} = kT.$$

Thus, the classical result (470) holds whenever the thermal energy greatly exceeds the typical spacing between quantum energy levels.

Consider the limit

$$\beta\hbar\omega\frac{\hbar\omega}{kT} >> 1,$$

in which the thermal energy is small compared to the separation between the energy levels. In this limit,

$$\exp(\beta\hbar\omega) >> 1,$$

Thus, if the thermal energy is much less than the spacing between quantum states then the mean energy approaches that of the ground-state (the so-called zero point energy). Clearly, the equipartition theorem is only valid in the former limit, where $kT >> \hbar\omega$, and the oscillator possess sufficient thermal energy to explore many of its possible quantum states.

SPECIFIC HEATS

We have discussed the internal energies and entropies of substances (mostly ideal gases) at some length. Unfortunately, these quantities cannot be directly measured. Instead, they must be inferred from other information. The thermodynamic property of substances which is the easiest to measure is, of course, the heat capacity, or specific heat. In fact, once the variation of the specific heat with temperature is known, both the internal energy and entropy can be easily reconstructed via

$$E(T, V) = v \int_0^T c_V(T, V) dT + E(0, V),$$

$$S(T, V) =$$

Here, use has been made of $dS = đQ/T$, and the third law of thermodynamics. Clearly, the optimum way of verifying the results of statistical thermodynamics is to compare the theoretically predicted heat capacities with the experimentally measured values.

Classical physics, in the guise of the equipartition theorem, says that each independent degree of freedom associated with a quadratic term in the energy possesses an average energy $(1/2)kT$ in thermal equilibrium at temperature T. Consider a substance made up of N molecules. Every molecular degree of freedom contributes $(1/2)N\,k\,T$, or $(1/2)\,vRT$, to the mean energy of the substance (with the tacit proviso that each degree of freedom is associated with a quadratic term in the energy). Thus, the contribution to the molar heat capacity at constant volume (we wish to avoid the complications associated with any external work done on the substance) is

$$\frac{1}{v}\left(\frac{\partial E}{\partial T}\right)_V = \frac{1}{v}\frac{\partial[(1/2)vRT]}{\partial T} = \frac{1}{2}R,$$

per molecular degree of freedom. The total classical heat capacity is therefore

$$c_V = \frac{g}{2}R,$$

where g is the number of molecular degrees of freedom. Since large complicated molecules clearly have very many more degrees of freedom than small simple molecules, the above formula predicts that the molar heat capacities of substances made up of the former type of molecules should greatly exceed those of substances made up of the latter. In fact, the experimental heat capacities of substances containing complicated molecules are generally greater than those of substances containing simple molecules, but by nowhere near the large factor predicted by Eq.

This equation also implies that heat capacities are temperature independent. In fact, this is not the case for most substances. Experimental heat capacities generally increase with increasing temperature. These two experimental facts pose severe problems for classical physics. Incidentally, these problems were fully appreciated as far back as 1850. Stories that physicists at the end of the nineteenth century thought that classical physics explained absolutely everything are largely apocryphal.

The equipartition theorem (and the whole classical approximation) is only valid when the typical thermal energy kT greatly exceeds the spacing between quantum energy levels. Suppose that the temperature is sufficiently low that this condition is not satisfied for one particular molecular degree of freedom. In fact, suppose that kT is much less than the spacing between the energy levels.

In this situation the degree of freedom only contributes the ground-state energy, E_0, say, to the mean energy of the molecule. The ground-state energy can be a quite complicated function of the internal properties of the molecule, but is certainly not a function of the temperature, since this is a collective property of all molecules. It follows that the contribution to the molar heat capacity is

$$\frac{1}{v}\left(\frac{\partial[N\,E_0]}{\partial T}\right)_v = 0.$$

Thus, if kT is much less than the spacing between the energy levels then the degree of freedom contributes nothing at all to the molar heat capacity. We say that this particular degree of freedom is frozen out. Clearly, at very low temperatures just about all degrees of freedom are frozen out. As the temperature is gradually increased, degrees of freedom successively "kick in," and eventually contribute their full (1/2)R to the molar heat capacity, as kT approaches, and then greatly exceeds, the spacing between their quantum energy levels. We can use these simple ideas to explain the behaviours of most experimental heat capacities.

To make further progress, we need to estimate the typical spacing between the quantum energy levels associated with various degrees of freedom. We can do this by observing the frequency of the electromagnetic radiation emitted and absorbed during transitions between these energy levels. If the typical spacing between energy levels is ΔE then transitions between the various levels are associated with photons of frequency v, where $hv = \Delta E$.

We can define an effective temperature of the radiation via $hv = kT_{rad}$. If $T \gg T_{rad}$ then $kT \gg \Delta E$, and the degree of freedom makes its full contribution to the heat capacity. On the other hand, if $T \ll T_{rad}$ then $kT \ll \Delta E$, and the degree of freedom is frozen out. The "temperatures" of various different types of radiation. It is clear that degrees of freedom which give rise to emission or absorption of radio or microwave radiation contribute their full $(1/2)R$ to the molar heat capacity at room temperature.

Degrees of freedom which give rise to emission or absorption in the visible, ultraviolet, X–ray, or γ–ray regions of the electromagnetic spectrum are frozen out at room temperature. Degrees of freedom which emit or absorb infrared radiation are on the border line.

Table. Effective "Temperatures" of Various Types of Electromagnetic Radiation

Radiation type	Frequency (Hz)	T_{rad} (°K)
Radio	$< 10^9$	< 0.05
Microwave	$10^9 - 10^{11}$	< 0.05
Infrared	$10^{11} - 10^{14}$	5. 5000
Visible	5×10^{14}	2×10^4
Ultraviolet	$10^{15} - 10^{17}$	5×10^4 5×10^5
X-ray	$10^{17} - 10^{20}$	5×10^6 5×10^9
g-ray	$> 10^{20}$	$> 5 \times 10^9 3$

Specific Heats of Gases

Let us now investigate the specific heats of gases. Consider, first of all, translational degrees of freedom. Every molecule in a gas is free to move in three dimensions. If one particular molecule has mass m and momentum $p = mv$ then its kinetic energy of translation is

$$K = \frac{1}{2m}\left(p_x^2 + p_y^2 + p_z^2\right).$$

The kinetic energy of other molecules does not involve the momentum P of this particular molecule. Moreover, the potential energy of interaction between molecules depends only on their position coördinates, and, thus, certainly does not involve P. Any internal rotational, vibrational, electronic, or nuclear degrees of freedom of the molecule also do not involve P.

Hence, the essential conditions of the equipartition theorem are satisfied (at least, in the classical approximation). Since Eq. contains three independent quadratic terms, there are clearly three degrees of freedom associated with translation (one for each dimension of space), so the translational contribution to the molar heat capacity of gases is

$$(c_V)_{\text{translation}} = \frac{3}{2}R.$$

Suppose that our gas is contained in a cubic enclosure of dimensions L. According to Schrödinger's equation, the quantized translational energy levels of an individual molecule are given by

$$E = \frac{\hbar^2 \pi^2}{2mL^2}\left(n_1^2 + n_2^2 + n_3^2\right),$$

where n_1, n_2, and n_3 are positive integer quantum numbers. Clearly, the spacing between the energy levels can be made arbitrarily small by increasing the size of the enclosure. This implies that translational degrees of freedom can be treated classically, so that Eq. is always valid (except very close to absolute zero). We conclude that all gases possess a minimum molar heat capacity of (3/2) R due to the translational degrees of freedom of their constituent molecules.

The electronic degrees of freedom of gas molecules (i.e., the possible configurations of electrons orbiting the atomic nuclei) typically give rise to absorption and emission in the ultraviolet or visible regions of the spectrum. It follows from electronic degrees of freedom are frozen out at room temperature.

Similarly, nuclear degrees of freedom (i.e., the possible configurations of protons and neutrons in the atomic nuclei) are frozen out because they are associated with absorption and emission in the X–ray and γ–ray regions of the electromagnetic spectrum. In fact, the only additional degrees of freedom we need worry about for gases are rotational and vibrational degrees of freedom. These typically give rise to absorption lines in the infrared region of the spectrum.

The rotational kinetic energy of a molecule tumbling in space can be written

$$K = \frac{1}{2} I_x \omega_x^2 + \frac{1}{2} I_y \omega_y^2 + \frac{1}{2} I_z \omega_z^2,$$

where the x –, y–, and z–axes are the so called principle axes of inertia of the molecule (these are mutually perpendicular), ω_x, ω_y, and ω_z are the angular velocities of rotation about these axes, and I_x, I_y, and I_z are the moments of inertia of the molecule about these axes. No other degrees of freedom depend on the angular velocities of rotation.

Since the kinetic energy of rotation is the sum of three quadratic terms, the rotational contribution to the molar heat capacity of gases is

$$(c_V)_{\text{rotation}} = \frac{3}{2}R,$$

according to the equipartition theorem. Note that the typical magnitude of a molecular moment of inertia is md^2, where m is the molecular mass, and d is the typical interatomic spacing in the molecule. A special case arises if the molecule is linear (e.g. if the molecule is diatomic).

In this case, one of the principle axes lies along the line of centers of the atoms. The moment of inertia about this axis is of order ma^2, where a is a typical nuclear dimension (remember that nearly all of the mass of an atom resides in the nucleus). Since $a \sim 10^{-5}\, d$,it follows that the moment of inertia about the line of centres is minuscule compared to the moments of inertia about the other two principle axes. In quantum mechanics, angular momentum is quantized in units of $\hbar$. The energy levels of a rigid rotator are written

$$E = \frac{\hbar^2}{2I} J(J+1),$$

where I is the moment of inertia and J is an integer. Note the inverse dependence of the spacing between energy levels on the moment of inertia.

It is clear that for the case of a linear molecule, the rotational degree of freedom associated with spinning along the line of centres of the atoms is frozen out at room temperature, given the very small moment of inertia along this axis, and, hence, the very widely spaced rotational energy levels.

Classically, the vibrational degrees of freedom of a molecule are studied by standard normal mode analysis of the molecular structure. Each normal mode behaves like an independent harmonic oscillator, and, therefore, contributes R to the molar specific heat of the gas [(1/2) R from the kinetic energy of vibration and (1/2)R from the potential energy of vibration].

A molecule containing n atoms has n–1 normal modes of vibration. For instance, a diatomic molecule has just one normal mode (corresponding to periodic stretching of the bond between the two atoms). Thus, the classical contribution to the specific heat from vibrational degrees of freedom is

$$(c_V)_{\text{vibration}} = (n-1)R.$$

The rotational and vibrational degrees of freedom actually make a contribution to the specific heats of gases at room temperature, once quantum effects are taken into consideration? We can answer this question by examining just one piece of data. The infrared absorption spectrum of Hydrogen Chloride. The absorption lines correspond to simultaneous transitions between different vibrational and rotational energy levels.

Hence, this is usually called a vibration-rotation spectrum. The missing line at about 3.47 microns corresponds to a pure vibrational transition from the ground-state to the first excited state (pure vibrational transitions are forbidden: HCl molecules always have to simultaneously change their rotational energy level if they are to couple effectively to electromagnetic radiation).

The longer wavelength absorption lines correspond to vibrational transitions in which there is a simultaneous decrease in the rotational energy level. Likewise, the shorter wavelength absorption lines correspond to vibrational transitions in which there is a simultaneous increase in the rotational energy level.

It is clear that the rotational energy levels are more closely spaced than the vibrational energy levels. The pure vibrational transition gives rise to absorption at about 3.47 microns, which corresponds to infrared radiation of frequency 8.5×10^{11} hertz with an associated radiation "temperature" of 4400 degrees kelvin.

We conclude that the vibrational degrees of freedom of HCl, or any other small molecule, are frozen out at room temperature. The rotational transitions split the vibrational lines by about 0.2 microns.

This implies that pure rotational transitions would be associated with infrared radiation of frequency 5×10^{12} hertz and corresponding radiation "temperature" 260 degrees kelvin. We conclude that the rotational degrees of freedom of

HCl, or any other small molecule, are not frozen out at room temperature, and probably contribute the classical (1/2)R to the molar specific heat. There is one proviso, however. Linear molecules (like HCl) effectively only have two rotational degrees of freedom (instead of the usual three), because of the very small moment of inertia of such molecules along the line of centres of the atoms.

We are now in a position to make some predictions regarding the specific heats of various gases. Monatomic molecules only possess three translational degrees of freedom, so monatomic gases should have a molar heat capacity (3/2)R = 12.47 joules/degree/mole.

The ratio of specific heats $\gamma = Cp/C_V = (C_V + R)/C_V$ should be 5/3 = 1.667. It can be seen from both of these predictions are borne out pretty well for Helium and Argon. Diatomic molecules possess three translational degrees of freedom and two rotational degrees of freedom (all other degrees of freedom are frozen out at room temperature). Thus, diatomic gases should have a molar heat capacity (5/2)R = 20.8 joules/degree/mole.

The ratio of specific heats should be 7/5 = 1.4 It can be seen from these are pretty accurate predictions for Nitrogen and Oxygen. The freezing out of vibrational degrees of freedom becomes gradually less effective as molecules become heavier and more complex.

This is partly because such molecules are generally less stable, so the force constant k is reduced, and partly because the molecular mass is increased. Both these effect reduce the frequency of vibration of the molecular normal modes and, hence, the spacing between vibrational energy levels.

This accounts for the obviously non-classical [i.e., not a multiple of (1/2)R] specific heats of Carbon Dioxide and Ethane. In both molecules, vibrational degrees of freedom contribute to the molar specific heat (but not the full R because the temperature is not high enough).

The variation of the molar heat capacity at constant volume (in units of R) of gaseous hydrogen with temperature. The expected contribution from the translational degrees of freedom is $(3/2)R$ (there are three translational degrees of freedom per molecule).

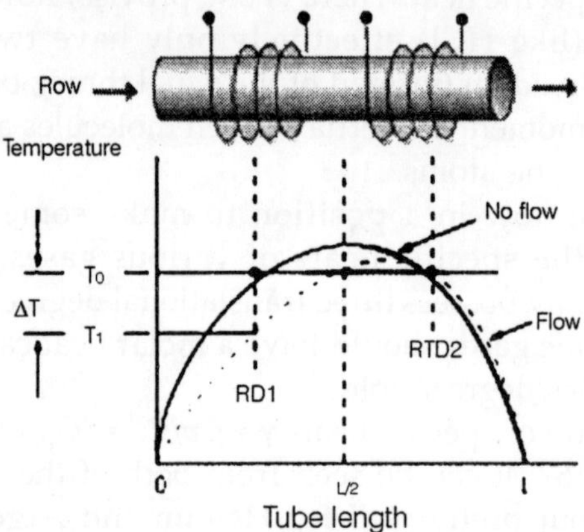

Fig. The Molar Heat Capacity at Constant Volume.

The expected contribution at high temperatures from the rotational degrees of freedom is R (there are effectively two rotational degrees of freedom per molecule). Finally, the expected contribution at high temperatures from the vibrational degrees of freedom is R (there is one vibrational degree of freedom per molecule). It can be seen that as the temperature rises the rotational, and then the vibrational, degrees of freedom eventually make their full classical contributions to the heat capacity.

Specific Heats of Solids

Consider a simple solid containing N atoms. Now, atoms in solids cannot translate (unlike those in gases), but are free to vibrate about their equilibrium positions. Such vibrations are called lattice vibrations, and can be thought of as sound waves propagating through the crystal lattice. Each atom is specified by three independent position coordinates, and three conjugate momentum coordinates.

Let us only consider small amplitude vibrations. In this case, we can expand the potential energy of interaction between the atoms to give an expression which is quadratic in the atomic displacements from their equilibrium positions.

It is always possible to perform a normal mode analysis of the oscillations. In effect, we can find 3*N* independent modes of oscillation of the solid.

Each mode has its own particular oscillation frequency, and its own particular pattern of atomic displacements. Any general oscillation can be written as a linear combination of these normal modes.

Let q_i be the (appropriately normalized) amplitude of the *i*th normal mode, and p_i the momentum conjugate to this coordinate. In normal mode coordinates, the total energy of the lattice vibrations takes the particularly simple form

$$E = \frac{1}{2}\sum_{i=1}^{3N}(p_i^2 + \omega_i^2 q_i^2),$$

where ω_i is the (angular) oscillation frequency of the *i*th normal mode. It is clear that in normal mode coordinates, the linearized lattice vibrations are equivalent to 3*N* independent harmonic oscillators (of course, each oscillator corresponds to a different normal mode).

The typical value of ω_i is the (angular) frequency of a sound wave propagating through the lattice. Sound wave frequencies are far lower than the typical vibration frequencies of gaseous molecules.

In the latter case, the mass involved in the vibration is simply that of the molecule, whereas in the former case the mass involved is that of very many atoms (since lattice vibrations are non-localized).

The strength of interatomic bonds in gaseous molecules is similar to those in solids, so we can use the estimate $\omega \sim \sqrt{k/m}$ (*k* is the force constant which measures the strength of interatomic bonds, and *m* is the mass involved in the oscillation) as proof that the typical frequencies of lattice vibrations are very much less than the vibration frequencies of simple molecules.

It follows from $\Delta E = \hbar\omega$ that the quantum energy levels of lattice vibrations are far more closely spaced than the vibrational energy levels of gaseous molecules. Thus, it is likely (and is, indeed, the case) that lattice vibrations are not frozen out at room temperature, but, instead, make their full classical contribution to the molar specific heat of the solid.

If the lattice vibrations behave classically then, according to the equipartition theorem, each normal mode of oscillation has an associated mean energy kT in equilibrium at temperature T [$(1/2)kT$ resides in the kinetic energy of the oscillation, and $(1/2)kT$ resides in the potential energy]. Thus, the mean internal energy per mole of the solid is

$$\bar{E} = 3NkT = 3vRT.$$

It follows that the molar heat capacity at constant volume is

$$c_V = \frac{1}{v}\left(\frac{\partial \bar{E}}{\partial T}\right)_V = 3R$$

for solids. This gives a value of 24.9 joules/mole/degree. In fact, at room temperature most solids (in particular, metals) have heat capacities which lie remarkably close to this value. This fact was discovered experimentally by Dulong and Petite at the beginning of the nineteenth century, and was used to make some of the first crude estimates of the molecular weights of solids (if we know the molar heat capacity of a substance then we can easily work out how much of it corresponds to one mole, and by weighing this amount, and then dividing the result by Avogadro's number, we can obtain an estimate of the molecular weight).

The experimental molar heat capacities Cp at constant pressure for various solids. The heat capacity at constant volume is somewhat less than the constant pressure value, but not by much, because solids are fairly incompressible. It can be seen that Dulong and Petite's law (i.e., that all solids have a molar heat capacities close to 24.9 joules/mole/degree) holds pretty well for metals. However, the law fails badly for diamond. This is not surprising.

As is well-known, diamond is an extremely hard substance, so its intermolecular bonds must be very strong, suggesting that the force constant k is large. Diamond is also a fairly low density substance, so the mass m involved in lattice vibrations is comparatively small. Both these facts suggest that the typical lattice vibration frequency of diamond ($\omega \sim \sqrt{k/m}$) is high. In fact, the spacing between the different vibration energy levels (which scales like $\hbar\omega$) is sufficiently large in diamond for the vibrational degrees of freedom to be largely frozen out at room temperature.

This accounts for the anomalously low heat capacity of diamond.

Table. Values of (Joules/Mole/Degree)
For Some Solids at T = 298° K. K. From Reif.

Solid	*Cp*	Solid	*Cp*
Copper	24.5	Aluminium	24.4
Silver	25.5	Tin (white)	26.4
Lead	26.4	Sulphur (rhombic)	22.4
Zinc	25.4	Carbon (diamond)	6.1

Dulong and Petite's law is essentially a high temperature limit. The molar heat capacity cannot remain a constant as the temperature approaches absolute zero, since, by Eq. this would imply $S \to \infty$, which violates the third law of thermodynamics. We can make a crude model of the behaviour of C_V at low temperatures by assuming that all the normal modes oscillate at the same frequency, ω, say.

According to Eq. the solid acts like a set of 3*N* independent oscillators which, making use of Einstein's approximation, all vibrate at the same frequency. We can use the quantum mechanical result for a single oscillator to write the mean energy of the solid in the form

$$\bar{E} = 3N\hbar\omega\left(\frac{1}{2} + \frac{1}{\exp(\beta\hbar\omega) - 1}\right).$$

The molar heat capacity is defined

$$c_V = \frac{1}{v}\left(\frac{\partial \bar{E}}{\partial T}\right)_V = \frac{1}{v}\left(\frac{\partial \bar{E}}{\partial T}\right)_V \frac{\partial \beta}{\partial T} = -\frac{1}{vkT^2}\left(\frac{\partial \bar{E}}{\partial T}\right)_V,$$

giving

$$c_V = -\frac{3\,N_A\hbar\omega}{kT^2}\left[-\frac{\exp(\beta\hbar\omega)}{[\exp(\beta\hbar\omega) - 1]^2}\right],$$

which reduces to

$$c_V = 3R\left(\frac{\theta_E}{T}\right)^2 \frac{\exp(\theta_E/T)}{[\exp(\theta_E/T)-1]^2}.$$

Here,

$$\theta_E = \frac{\hbar\omega}{k}$$

is called the Einstein temperature. If the temperature is sufficiently high that $T \gg \theta_E$ then $kT \gg \hbar\omega$, and the above expression reduces to $C_V = 3R$,, after expansion of the exponential functions. Thus, the law of Dulong and Petite is recovered for temperatures significantly in excess of the Einstein temperature. On the other hand, if the temperature is sufficiently low that $T \ll \theta_E$ then the exponential factors in Eq. become very much larger than unity, giving

$$C_V \sim 3R\left(\frac{\theta_E}{T}\right)\exp(-\theta_E/T).$$

So, in this simple model the specific heat approaches zero exponentially as $T \to 0$.

In reality, the specific heats of solids do not approach zero quite as quickly as suggested by Einstein's model when $T \to 0$. The experimentally observed low temperature behaviour is more like $C_V \propto T^3$. The reason for this discrepancy is the crude approximation that all normal modes have the same frequency. In fact, long wavelength modes have lower frequencies than short wavelength modes, so the former are much harder to freeze out than the latter (because the spacing between quantum energy levels, $\hbar\omega$, is smaller in the former case).

The molar heat capacity does not decrease with temperature as rapidly as suggested by Einstein's model because these long wavelength modes are able to make a significant contribution to the heat capacity even at very low temperatures.

A more realistic model of lattice vibrations was developed by the Dutch physicist Peter Debye in 1912. In the Debye model, the frequencies of the normal modes of vibration are estimated by treating the solid as an isotropic continuous medium. This approach is reasonable because the only modes which really matter at low temperatures are the long wavelength modes: i.e., those whose wavelengths greatly exceed the interatomic spacing. It is plausible that these modes are not particularly sensitive to the discrete nature of the solid: i.e., the fact that it is made up of atoms rather than being continuous.

Consider a sound wave propagating through an isotropic continuous medium. The disturbance varies with position vector *r* and time *t* like $\exp[-i\,(\mathrm{k.r} - \omega t)]$, where the wave-vector *k* and the frequency of oscillation ω satisfy the dispersion relation for sound waves in an isotropic medium:

$$\omega = kC_s.$$

Here, C_S is the speed of sound in the medium. Suppose, for the sake of argument, that the medium is periodic in the *x*–, *y*–, and *z*–directions with periodicity lengths L_x, L_y, and L_z, respectively. In order to maintain periodicity we need

$$k_x(x + L_x) = k_x + 2\pi n_x,$$

where n_x is an integer. There are analogous constraints on k_y and k_z. It follows that in a periodic medium the components of the wave-vector are quantized, and can only take the values

$$k_x = \frac{2\pi}{L_x} n_x,$$

$$k_y = \frac{2\pi}{L_y} n_y,$$

$$k_z = \frac{2\pi}{L_z} n_z,$$

where n_x, n_y, and n_z are all integers. It is assumed that L_x, L_y, and L_z are macroscopic lengths, so the allowed values of the components of the wave-vector are very closely spaced. For given values of k_y and k_z, the number of allowed values of k_x which lie in the range k_x to $k_x + dk_x$ is given by

$$\Delta n_x = \frac{L_x}{2\pi} 2k_x.$$

It follows that the number of allowed values of k (i.e., the number of allowed modes) when k_x lies in the range k_z to $k_z + dk_z$, k_z, $+ k_y$ lies in the range k_y to $k_y + dk_y$, and k_z lies in the range k_z to $k_z + dk_z$, is

$$\rho d^3k = \left(\frac{L_x}{2\pi}dk_x\right)\left(\frac{L_y}{2\pi}dk_y\right)\left(\frac{L_z}{2\pi}dk_z\right) = \frac{V}{(2\pi)^3}dk_x dk_y dk_z,$$

where $V = L_x L_y L_z$ is the periodicity volume, and $d^3k \equiv dk_x dk_y dk_z$. The quantity ρ is called the density of modes. Note that this density is independent of k, and proportional to the periodicity volume.

Thus, the density of modes per unit volume is a constant independent of the magnitude or shape of the periodicity volume. The density of modes per unit volume when the magnitude of k lies in the range k to $k + dk$ is given by multiplying the density of modes per unit volume by the "volume" in k-space of the spherical shell lying between radii k and $k + dk$. Thus,

$$\rho_k dk = \frac{4\pi k^2\, dk}{(2\pi)} = \frac{V^2}{2\pi^2}dk.$$

Consider an isotropic continuous medium of volume V. According to the above relation, the number of normal modes whose frequencies lie between ω and $\omega + d\omega$ (which is equivalent to the number of modes whose k values lie in the range ω/C_s to $\omega/C_s + d\omega/C_s$) is

$$\sigma_C(\omega)\, d\omega = 3\frac{k^2\, V}{2\pi^2}dk = 3\frac{V^2}{2\pi^2 C_s{}^3}\omega^2 d\omega.$$

The factor of 3 comes from the three possible polarizations of sound waves in solids. For every allowed wavenumber (or frequency) there are two independent torsional modes, where the displacement is perpendicular to the direction of propagation, and one longitudinal mode, where the displacement is parallel to the direction of propagation. Torsion waves are vaguely analogous to electromagnetic waves (these also have two independent polarizations). The longitudinal mode is very similar to the compressional sound wave in gases. Of course, torsion waves can not propagate in gases because gases have no resistance to deformation without change of volume.

The Debye approach consists in approximating the actual density of normal modes $\sigma(\omega)$ by the density in a continuous medium $\sigma_C(\omega)$, not only at low frequencies (long wavelengths) where these should be nearly the same, but also at higher frequencies where they may differ substantially.

Suppose that we are dealing with a solid consisting of N atoms. We know that there are only $3N$ independent normal modes. It follows that we must cut off the density of states above some critical frequency, ω_D say, otherwise we will have too many modes. Thus, in the Debye approximation the density of normal modes takes the form

$$\sigma_D(\omega) = \sigma_C(\omega) \text{ for } \omega < \omega_D$$
$$\sigma_D(\omega) = 0 \text{ for } \omega < \omega_D$$

Here, ω_D is the Debye frequency. This critical frequency is chosen such that the total number of normal modes is $3N$, so

$$\int_0^{\infty} \sigma_C(\omega)\, d\omega = \int_0^{\omega_D} \sigma_C(\omega)\, d\omega = 3N.$$

Substituting Eq. into the previous formula yields

$$\frac{3V}{2\pi^2 C_s^3} \int_0^{\omega_D} d\omega = \frac{3V}{2\pi^2 C_s^3} \omega_D^3 = 3N.$$

This implies that

$$\omega_D = C_s \left(6\pi^2 \frac{N}{V} \right)^{1/3}.$$

Thus, the Debye frequency depends only on the sound velocity in the solid and the number of atoms per unit volume. The wavelength corresponding to the Debye frequency is $2\pi C_s/\omega_D$, which is clearly on the order of the interatomic spacing $a \sim (V/N)^{1/3}$.

It follows that the cut-off of normal modes whose frequencies exceed the Debye frequency is equivalent to a cut-off of normal modes whose wavelengths are less than the interatomic spacing. Of course, it makes physical sense that such modes should be absent.

Compares the actual density of normal modes in diamond with the density predicted by Debye theory. Not surprisingly, there is not a particularly strong resemblance between these two curves, since Debye theory is highly idealized. Nevertheless, both curves exhibit sharp cut-offs at high frequencies, and coincide at low frequencies. Furthermore, the areas under both curves are the same. This is sufficient to allow Debye theory to correctly account for the temperature variation of the specific heat of solids at low temperatures.

We can use the quantum mechanical expression for the mean energy of a single oscillator, Eq. to calculate the mean energy of lattice vibrations in the Debye approximation. We obtain

$$\bar{E} = \int_0^\infty \sigma_D(\omega)\hbar\omega\left(\frac{1}{2}+\frac{1}{\exp(\beta\hbar\omega)-1}\right)d\omega.$$

According to Eq. the molar heat capacity takes the form

$$C_V = \frac{1}{\upsilon kT^2}\int_0^\infty \sigma_D(\omega)\hbar\omega\left[\frac{\exp(\beta\hbar\omega)\hbar\omega}{[\exp(\beta\hbar\omega)-1]^2}\right]d\omega.$$

Substituting in Eq. we find that

$$C_V = \frac{k}{\upsilon}\int_0^{\omega_D}\frac{\exp(\beta\hbar\omega)(\beta\hbar\omega)^2}{[\exp(\beta\hbar\omega)-1]^2}\frac{3V}{2\pi^2 C_s^3}\omega^2 d\omega,$$

giving

$$C_V = \frac{3Vk}{2\pi^2\upsilon(C_s\beta\hbar)^3}\int_0^{\beta\hbar\omega_D}\frac{\exp x}{(\exp x-1)^2}x^4 dx,$$

in terms of the dimensionless variable $x = \beta\hbar\omega$. According to Eq. the volume can be written

$$V = 6\pi^2 N\left(\frac{Cs}{\omega_D}\right)^3,$$

so the heat capacity reduces to

$$C_V = 3Rf_D(\beta\hbar\omega_D) = 3Rf_D(\theta_D/T),$$

where the Debye function is defined

$$f_D(y) \equiv \frac{3}{y^3}\int_0^y \frac{\exp x}{(\exp x - 1)^2} x^4 dx.$$

We have also defined the Debye temperature θ_D as

$$k\theta_D = \hbar\omega_D.$$

Consider the asymptotic limit in which $T \gg \theta_D$. For small y, we can approximate $\exp x$ as $1 + x$ in the integrand of Eq. so that

$$f_D(y) \to \frac{3}{y^3}\int_0^y x^2 dx = 1.$$

Thus, if the temperature greatly exceeds the Debye temperature we recover the law of Dulong and Petite that $C_V = 3R$. Consider, now, the asymptotic limit in which $T \ll \theta_D$. For large y,

$$\int_0^y \frac{\exp x}{(\exp x - 1)^2} x^4 dx \simeq \int_0^\infty \frac{\exp x}{(\exp x - 1)^2} x^4 dx = \frac{4\pi^4}{15}.$$

Thus, in the low temperature limit

$$f_D(y) \to \frac{4\pi^4}{5}\frac{1}{y^3}.$$

This yields

$$c_V \simeq \frac{12\pi^4}{5} R \left(\frac{T}{\theta_D}\right)^3$$

in the limit $T \ll \theta_D$: i.e., c_V varies with temperature like T^3.

Table. Comparison of Debye Temperatures (In Degrees Kelvin)

Solid	θ_D *from low temp.*	θ_D *from sound speed*
NaCl	308	320
KCl	230	246
Ag	225	216
Zn	308	305

The fact that c_V goes like T^3 at low temperatures is quite well verified experimentally, although it is sometimes necessary to go to temperatures as low as 0.02 θ_D to obtain this asymptotic behaviour.

Theoretically, θ_D should be calculable from Eq. in terms of the sound speed in the solid and the molar volume. A comparison of Debye temperatures evaluated by this means with temperatures obtained empirically by fitting the law to the low temperature variation of the heat capacity. It can be seen that there is fairly good agreement between the theoretical and empirical Debye temperatures. This suggests that the Debye theory affords a good, thought not perfect, representation of the behaviour of c_V in solids over the entire temperature range.

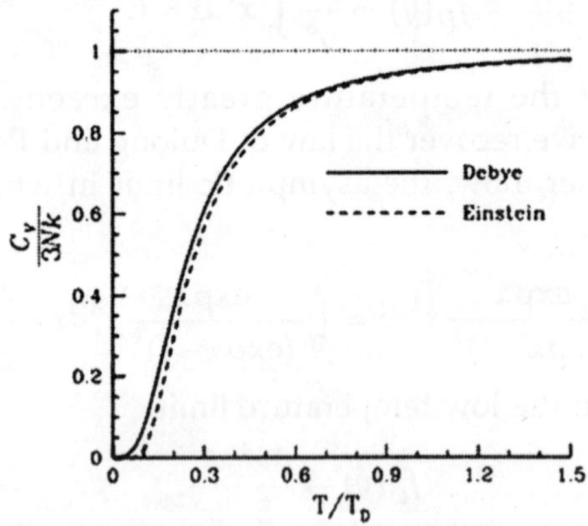

Fig. The Molar Heat Capacity of Various Solids.

Finally, the actual temperature variation of the molar heat capacities of various solids as well as that predicted by Debye's theory. The prediction of Einstein's theory is also show for the sake of comparison. Note that 24.9 joules/mole/degree is about 6 calories/gram-atom/degree (the latter are chemist's units).

THE MAXWELL DISTRIBUTION

Consider a molecule of mass m in a gas which is sufficiently dilute for the intermolecular forces to be negligible (i.e., an ideal gas). The energy of the molecule is written

$$\epsilon = \frac{P^2}{2m} + \epsilon^{\text{int}},$$

where p is its momentum vector, and $\in^{int}$ is its internal (i.e., non-translational) energy. The latter energy is due to molecular rotation, vibration, etc.

Translational degrees of freedom can be treated classically to an excellent approximation, whereas internal degrees of freedom usually require a quantum mechanical approach.

Classically, the probability of finding the molecule in a given internal state with a position vector in the range r to $r + dr$, and a momentum vector in the range P to $P + dP$, is proportional to the number of cells (of "volume" h_0) contained in the corresponding region of phase-space, weighted by the Boltzmann factor.

In fact, since classical phase-space is divided up into uniform cells, the number of cells is just proportional to the "volume" of the region under consideration. This "volume" is written d^3rd^3p. Thus, the probability of finding the molecule in a given internal state s is

$$P_s(r,p)d^3r\,d^3p \propto \exp(-\beta p^2/2m)\exp(\beta \in_s^{int})d^3r\,d^3p,$$

where Ps is a probability density defined in the usual manner. The probability $P(r, p)\ d^3r\ d^3p$ of finding the molecule in any internal state with position and momentum vectors in the specified range is obtained by summing the above expression over all possible internal states.

The sum over $\exp(-\beta \in_s^{int})$ just contributes a constant of proportionality (since the internal states do not depend on r or p), so

$$P(r,p)d^3r\,d^3p \propto \exp(-\beta p^2/2m)\,d^3r\,d^3p.$$

Of course, we can multiply this probability by the total number of molecules N in order to obtain the mean number of molecules with position and momentum vectors in the specified range.

Suppose that we now want to determine:$f(r, v)\ d^3r\ d^3v$ i.e., the mean number of molecules with positions between r and $r + dr$, and velocities in the range v and $v + dv$. Since $v = p/m$, it is easily seen that

$$f(r,v)d^3r\,d^3v = C\exp(-\beta mv^2/2)\,d^3r\,d^3v,$$

where C is a constant of proportionality. This constant can be determined by the condition

$$\int_{(r)}\int_{(v)} f(r,v)d^3r\,d^3v = N:$$

i.e., the sum over molecules with all possible positions and velocities gives the total number of molecules, N. The integral over the molecular position coordinates just gives the volume V of the gas, since the Boltzmann factor is independent of position. The integration over the velocity coordinates can be reduced to the product of three identical integrals (one for v_x, one for v_y, and one for v_z), so we have

$$CV\left[\int_{-\infty}^{\infty} \exp(-\beta m v_z^2/2)dv_z\right]^3 = N.$$

Now,

$$\int_{-\infty}^{\infty} \exp(-\beta m v_z^2/2)dv_z = \sqrt{\frac{2}{\beta m}}\int_{-\infty}^{\infty}\exp(-y^2)dy = \sqrt{\frac{2\pi}{\beta m}},$$

so $C = (N/V)\,(\beta m/2\pi)^{3/2}$. Thus, the properly normalized distribution function for molecular velocities is written

$$f(v)d^3r\,d^3v = n\left(\frac{m}{2\pi kT}\right)^{3/2}\exp(-mv^2/2kT)d^3r\,d^3v.$$

Here, $n = N/V$ is the number density of the molecules. We have omitted the variable r in the argument of f, since f clearly does not depend on position. In other words, the distribution of molecular velocities is uniform in space. This is hardly surprising, since there is nothing to distinguish one region of space from another in our calculation.

The above distribution is called the Maxwell velocity distribution, because it was discovered by James Clark Maxwell in the middle of the nineteenth century. The average number of molecules per unit volume with velocities in the range v to $v + dv$ is obviously $f(v)\,d^3v$.

Let us consider the distribution of a given component of velocity: the z-component, say. Suppose that $g(v_z)\,dv_z$ is the average number of molecules per unit volume with the z-component of velocity in the range v_z to $v_z + dv_z$, irrespective of the values of their other velocity components.

It is fairly obvious that this distribution is obtained from the Maxwell distribution by summing (integrating actually) over all possible values of v_x and v_y, with v_z in the specified range. Thus,

$$g(vz)dv_z = \int_{(v_z)}\int_{(v_y)} f(v)d^3v.$$

This gives

$$g(v_z)dv_z = n\left(\frac{m}{2\pi kT}\right)$$

$$\int_{(v_z)}\int_{(v_y)} \exp\left[-(m/2kT)\left(v_z^2+v_v^2+v_z^2\right)\right]dv_x dv_y dv_z$$

$$= n\left(\frac{m}{2\pi\, kT}\right)^{3/2} \exp\left(-mv_z^2/2kT\right)\left[\int_{-\infty}^{\infty}\exp\left(-mv_x^2/2k\,T\right)\right]^2$$

$$= n\left(\frac{m}{2\pi\, kT}\right)^{3/2} \exp\left(-mv_z^2/2\,kT\right)\left(\sqrt{\frac{2\pi\, kT}{m}}\right)^2,$$

or

$$g(v_z)\,dv_z = n\left(\frac{m}{2\pi\, kT}\right)^{1/2} \exp\left(-mv_z^2/2\,kT\right)dv_z.$$

Of course, this expression is properly normalized, so that

$$\int_{-\infty}^{\infty} g(v_z)dv_z = n.$$

It is clear that each component (since there is nothing special about the z-component) of the velocity is distributed with a Gaussian probability distribution, centred on a mean value

$$\overline{v_z} = 0,$$

with variance

$$\overline{v_z^2} = \frac{kT}{m}.$$

Equation implies that each molecule is just as likely to be moving in the plus z-direction as in the minus z-direction. Equation can be rearranged to give

$$\overline{\frac{1}{2}mv_z^2} = \frac{1}{2}kT,$$

in accordance with the equipartition theorem.

Note that Eq. can be rewritten

$$\frac{f(v)d^3v}{n} = \left[\frac{g(v_x)dv_x}{n}\right]\left[\frac{g(v_y)dv_y}{n}\right]\left[\frac{g(v_z)dv_x}{n}\right],$$

where $g(v_x)$ and $g(v_y)$ are defined in an analogous way to $g(v_z)$. Thus, the probability that the velocity lies in the range v to $v + dv$ is just equal to the product of the probabilities that the velocity components lie in their respective ranges. In other words, the individual velocity components act like statistically independent variables.

Suppose that we now want to calculate $F(v)\ dv$: i.e., the average number of molecules per unit volume with a speed $v = |v|$ in the range v to $v + dv$. It is obvious that we can obtain this quantity by adding up all molecules with speeds in this range, irrespective of the direction of their velocities. Thus,

$$F(v)dv = \int f(v)d^3v,$$

where the integral extends over all velocities satisfying

$$v < |v| < v + dv.$$

This inequality is satisfied by a spherical shell of radius v and thickness dv in velocity space. Since $f(v)$ only depends on $|v|$, so $f(v) \equiv f(v)$, the above integral is just $f(v)$ multiplied by the volume of the spherical shell in velocity space. So,

$$F(v)dv = 4\pi f(v)v^2dv,$$

which gives

$$F(v)dv = 4\pi f(v)\left(\frac{m}{2\pi\, kT}\right)v^2 \exp(-mv^2/2k\,T)\,dv.$$

This is the famous Maxwell distribution of molecular speeds. Of course, it is properly normalized, so that

$$\int_0^\infty F(v)dv = n.$$

Note that the Maxwell distribution exhibits a maximum at some non-zero value of v. The reason for this is quite simple. As v increases, the Boltzmann factor decreases, but the volume of phase-space available to the molecule (which is proportional to v^2) increases: the net result is a distribution with a non-zero maximum.

The mean molecular speed is given by

$$\bar{v} = \frac{1}{n}\int_0^\infty F(v)v\,dv.$$

Thus, we obtain

$$\bar{v} = 4\pi\left(\frac{m}{2\pi kT}\right)^{3/2}\int_0^\infty v^3 \exp(-mv^2/2kT)\,dv,$$

or

$$\bar{v} = 4\pi\left(\frac{m}{2\pi kT}\right)^{3/2}\left(\frac{2kT}{m}\right)^2\int_0^\infty y^3 \exp(-y^2)\,dy$$

Now

$$\int_0^\infty y^3 \exp(-y^2)\,dy = \frac{1}{2},$$

so

$$\bar{v} = \sqrt{\frac{8}{\pi}\frac{kT}{m}}.$$

A similar calculation gives

$$v_{\rm rms} = \sqrt{\overline{v^2}} = \sqrt{\frac{3\,kT}{m}}.$$

However, this result can also be obtained from the equipartition theorem. Since

$$\overline{\frac{1}{2}mv^2} = \overline{\frac{1}{2}m\left(v_z^2 + v_y^2 + v_z^2\right)} = 3\left(\frac{1}{2}kT\right),$$

then Eq. follows immediately. It is easily demonstrated that the most probable molecular speed (i.e., the maximum of the Maxwell distribution function) is

$$\overline{v} = \sqrt{\frac{2kT}{m}}.$$

The speed of sound in an ideal gas is given by

$$c_s = \sqrt{\frac{\gamma\, kT}{m}},$$

where γ is the ratio of specific heats. This can also be written since $p = nkT$ and $\rho = nm$. It is clear that the various average speeds which we have just calculated are all of order the sound speed (i.e., a few hundred meters per second at room temperature). In ordinary air ($\gamma = 1.4$) the sound speed is about 84% of the most probable molecular speed, and about 74% of the mean molecular speed. Since sound waves ultimately propagate via molecular motion, it makes sense that they travel at slightly less than the most probable and mean molecular speeds.

The Maxwell velocity distribution as a function of molecular speed in units of the most probable speed. Also shown are the mean speed and the root mean square speed.

It is difficult to directly verify the Maxwell velocity distribution. However, this distribution can be verified indirectly by measuring the velocity distribution of atoms exiting from a small hole in an oven.

The velocity distribution of the escaping atoms is closely related to, but slightly different from, the velocity distribution inside the oven, since high velocity atoms escape more readily than low velocity atoms. In fact, the predicted velocity distribution of the escaping atoms varies like $v^3 \exp(-mv^2/2kT)$, in contrast to the $v^3 \exp(-mv^2/2kT)$ variation of the velocity distribution inside the oven. The measured and theoretically predicted velocity distributions of potassium atoms escaping from an oven at 157°C. There is clearly very good agreement between the two.

Chapter 5

Classical Thermodynamics

We have learned that macroscopic quantities such as energy, temperature, and pressure are, in fact, statistical in nature: i.e., in equilibrium they exhibit random fluctuations about some mean value. If we were to plot out the probability distribution for the energy, say, of a system in thermal equilibrium with its surroundings we would obtain a Gaussian with a very small fractional width. In fact, we expect

$$\frac{\Delta^* E}{\overline{E}} \sim \frac{1}{\sqrt{f}},$$

where the number of degrees of freedom f is about 10^{24} for laboratory scale systems. This means that the statistical fluctuations of macroscopic quantities about their mean values are typically only about 1 in 10^{12}.

Since the statistical fluctuations of equilibrium quantities are so small, we can neglect them to an excellent approximation, and replace macroscopic quantities, such as energy, temperature, and pressure, by their mean values. So, $p \to \overline{p}$, and $T \to \overline{T}$, etc. In the following discussion, we shall drop the overbars altogether, so that p should be understood to represent the mean pressure $\overline{p}$, etc. This prescription, which is the essence of classical thermodynamics, is equivalent to replacing all statistically varying quantities by their most probable values.

Although there are formally four laws of thermodynamics (i.e., the zeroth to the third), the zeroth law is really a consequence of the second law, and the third law is actually only important at temperatures close to absolute zero. So, for most purposes, the two laws which really matter are the first law and the second law. For an infinitesimal process, the first law is written

$$đQ = dE + đW,$$

where dE is the change in internal energy of the system, $đQ$ is the heat absorbed by the system, and $đW$ is the work done by the system on its surroundings. Note that this is just a convention. We could equally well write the first law in terms of the heat emitted by the system or the work done on the system. It does not really matter, as long as we are consistent in our definitions.

The second law of thermodynamics implies that

$$đQ = TdS,$$

for a quasi-static process, where T is the thermodynamic temperature, and dS is the change in entropy of the system. Furthermore, for systems in which the only external parameter is the volume (i.e., gases), the work done on the environment is

$$đW = pdV,$$

where p is the pressure, and dV is the change in volume. Thus, it follows from the first and second laws of thermodynamics that

$$T\,dS = dE + p\,dV.$$

THE EQUATION OF STATE OF AN IDEAL GAS

Let us start our discussion by considering the simplest possible macroscopic system: i.e., an ideal gas. All of the thermodynamic properties of an ideal gas are summed up in its equation of state, which determines the relationship between its pressure, volume, and temperature. Unfortunately, classical thermodynamics is unable to tell us what this equation of state is from first principles. In fact, classical thermodynamics cannot tell us anything from first principles. We always have to provide some information to begin with before classical thermodynamics can generate any new results. This initial information may come from statistical physics (i.e., from our knowledge of the microscopic structure of the system under consideration), but, more usually, it is entirely empirical in nature (i.e., it is the result of experiments). Of course, the ideal gas law was first discovered empirically by Robert Boyle, but, nowadays, we can justify it from statistical arguments. Recall, that the number of accessible states of a monotonic ideal gas varies like

$$\Omega \propto V^N \chi(E),$$

where N is the number of atoms, and $\chi(E)$ depends only on the energy of the gas (and is independent of the volume). We obtained this result by integrating over the volume of accessible phase-space. Since the energy of an ideal gas is independent of the particle coordinates (because there are no interatomic forces), the integrals over the coordinates just reduced to N simultaneous volume integrals, giving the V^N factor in the above expression. The integrals over the particle momenta were more complicated, but were clearly completely independent of V, giving the $\chi(E)$ factor in the above expression. Now, we have a statistical rule which tells us that

$$X_a - \frac{1}{\beta}\frac{\partial \ln \Omega}{\partial x_a},$$

where X_a is the mean force conjugate to the external parameter x_a (i.e., $đW = \sum_a X_a dx_a$), and $\beta = 1/kT$. For an ideal gas, the only external parameter is the volume, and its conjugate force is the pressure (since $đW = pdV$). So, we can write

$$p = \frac{1}{\beta}\frac{\partial \ln \Omega}{\partial V}.$$

If we simply apply this rule to Eq. we obtain

$$p = \frac{N\,k\,T}{V}.$$

However, $N = v\,N_A$, where v is the number of moles, and N_A is Avagadro's number. Also, $k\,N_A = R$, where R is the ideal gas constant. This allows us to write the equation of state in its usual form

$$pV = vRT.$$

The above derivation of the ideal gas equation of state is rather elegant. It is certainly far easier to obtain the equation of state in this manner than to treat the atoms which make up the gas as little billiard balls which continually bounce of the walls of a container.

The latter derivation is difficult to perform correctly because it is necessary to average over all possible directions of atomic motion.

It is clear, from the above derivation, that the crucial element needed to obtain the ideal gas equation of state is the absence of interatomic forces. This automatically gives rise to a variation of the number of accessible states with E and V of the form which, in turn, implies the ideal gas law. So, the ideal gas law should also apply to polyatomic gases with no interatomic forces.

Polyatomic gases are more complicated that monatomic gases because the molecules can rotate and vibrate, giving rise to extra degrees of freedom, in addition to the translational degrees of freedom of a monatomic gas. In other words, $\chi(E)$, in Eq. becomes a lot more complicated in polyatomic gases. However, as long as there are no interatomic forces, the volume dependence of Ω is still V^N, and the ideal gas law should still hold true. In fact, we shall discover that the extra degrees of freedom of polyatomic gases manifest themselves by increasing the specific heat capacity.

There is one other conclusion we can draw from Eq. The statistical definition of temperature is

$$\frac{1}{kT} = \frac{\partial \ln \Omega}{\partial E}.$$

It follows that

$$\frac{1}{kT} = \frac{\partial \ln \chi}{\partial E}.$$

We can see that since χ is a function of the energy, but not the volume, then the temperature must be a function of the energy, but not the volume. We can turn this around and write

$$E = E(T).$$

In other words, the internal energy of an ideal gas depends only on the temperature of the gas, and is independent of the volume. This is pretty obvious, since if there are no interatomic forces then increasing the volume, which effectively increases the mean separation between molecules, is not going to affect the molecular energies in any way. Hence, the energy of the whole gas is unaffected.

The volume independence of the internal energy can also be obtained directly from the ideal gas equation of state. The internal energy of a gas can be considered as a general function of the temperature and volume, so

$$E = E(T, V).$$

It follows from mathematics that

$$dE = \left(\frac{\partial E}{\partial T}\right)_V dT + \left(\frac{\partial E}{\partial T}\right)_T dV,$$

where the subscript V reminds us that the first partial derivative is taken at constant volume, and the subscript T reminds us that the second partial derivative is taken at constant temperature. Thermodynamics tells us that for a quasi-static change of parameters

$$T\,dS = dE + pdV.$$

The ideal gas law can be used to express the pressure in term of the volume and the temperature in the above expression. Thus,

$$dS = \frac{1}{T}dE + \frac{\upsilon R}{V}dV.$$

Using Eq. this becomes

$$dS = \frac{1}{T}\left(\frac{\partial E}{\partial T}\right)_V dT + \left[\frac{1}{T}\left(\frac{\partial E}{\partial V}\right)_T + \frac{\upsilon R}{V}\right]dV.$$

However, dS is the exact differential of a well-defined state function, S. This means that we can consider the entropy to be a function of temperature and volume. Thus, $S = S(T, V)$, and mathematics immediately tells us that

$$dS = \left(\frac{\partial E}{\partial T}\right)_V dT + \left(\frac{\partial E}{\partial V}\right)_T dV.$$

The above expression is true for all small values of dT and dV, so a comparison with Eq. gives

$$\left(\frac{\partial S}{\partial T}\right)_V = \frac{1}{T}\left(\frac{\partial E}{\partial T}\right)_V,$$

$$\left(\frac{\partial S}{\partial T}\right)_T = \frac{1}{T}\left(\frac{\partial E}{\partial V}\right)_T + \frac{vR}{V}.$$

One well-known property of partial differentials is the equality of second derivatives, irrespective of the order of differentiation, so

$$\frac{\partial^2 S}{\partial V\,\partial T} = \frac{\partial^2 S}{\partial T\,\partial V}.$$

This implies that

$$\left(\frac{\partial}{\partial V}\right)_T\left(\frac{\partial S}{\partial T}\right)_V = \left(\frac{\partial}{\partial T}\right)_V\left(\frac{\partial S}{\partial T}\right)_T.$$

The above expression can be combined with Eqs. to give

$$\frac{1}{T}\left(\frac{\partial^2 E}{\partial V\,\partial T}\right) = \left[\frac{1}{T^2}\left(\frac{\partial E}{\partial V}\right)_T + \frac{1}{T}\left(\frac{\partial^2 E}{\partial T\,\partial V}\right)\right].$$

Since second derivatives are equivalent, irrespective of the order of differentiation, the above relation reduces to

$$\left(\frac{\partial E}{\partial V}\right)_T = 0,$$

which implies that the internal energy is independent of the volume for any gas obeying the ideal equation of state. This result was confirmed experimentally by James Joule in the middle of the nineteenth century.

HEAT CAPACITY OR SPECIFIC HEAT

Suppose that a body absorbs an amount of heat ΔQ and its temperature consequently rises by ΔT. The usual definition of the heat capacity, or specific heat, of the body is

$$C = \frac{\Delta Q}{\Delta T}.$$

If the body consists of v moles of some substance then the molar specific heat (i.e., the specific heat of one mole of this substance) is defined

$$c = \frac{1}{v}\frac{\Delta Q}{\Delta T}.$$

In writing the above expressions, we have tacitly assumed that the specific heat of a body is independent of its temperature. In general, this is not true. We can overcome this problem by only allowing the body in question to absorb a very small amount of heat, so that its temperature only rises slightly, and its specific heat remains approximately constant. In the limit as the amount of absorbed heat becomes infinitesimal, we obtain

$$c = \frac{1}{v}\frac{đW}{dT}.$$

In classical thermodynamics, it is usual to define two specific heats. Firstly, the molar specific heat at constant volume, denoted

$$c_p = \frac{1}{v}\left(\frac{đQ}{dT}\right)_V.$$

and, secondly, the molar specific heat at constant pressure, denoted

$$c_p = \frac{1}{v}\left(\frac{đQ}{dT}\right)_p.$$

Consider the molar specific heat at constant volume of an ideal gas. Since $dV = 0$, no work is done by the gas, and the first law of thermodynamics reduces to

$$đQ = dE.$$

It follows from Eq. that

$$c_V = \frac{1}{v}\left(\frac{\partial E}{\partial T}\right)_V.$$

Now, for an ideal gas the internal energy is volume independent. Thus, the above expression implies that the specific heat at constant volume is also volume independent. Since E is a function only of T, we can write

$$dE = \left(\frac{\partial E}{\partial T}\right)_V dT.$$

The previous two expressions can be combined to give

$$dE = v\, c_V\, dT$$

for an ideal gas.

Let us now consider the molar specific heat at constant pressure of an ideal gas. In general, if the pressure is kept constant then the volume changes, and so the gas does work on its environment. According to the first law of thermodynamics,

$$đQ = dE + pdV = vc_V\, dT + pdV.$$

The equation of state of an ideal gas tells us that if the volume changes by dV, the temperature changes by dT, and the pressure remains constant, then

$$pdV = vRdT.$$

The previous two equations can be combined to give

$$đQ = vVdT + vRdT.$$

Now, by definition

$$c_p = \frac{1}{v}\left(\frac{đQ}{dT}\right)_p,$$

so we obtain

$$c_p = c_V + R.$$

for an ideal gas. This is a very famous result. Note that at constant volume all of the heat absorbed by the gas goes into increasing its internal energy, and, hence, its temperature, whereas at constant pressure some of the absorbed heat is used to do work on the environment as the volume increases. This means that, in the latter case, less heat is available to increase the temperature of the gas. Thus, we expect the specific heat at constant pressure to exceed that at constant volume, as indicated by the above formula.

The ratio of the two specific heats Cp/C_V is conventionally denoted γ. We have

$$\gamma \equiv \frac{c_p}{c_V} = 1 + \frac{R}{c_V}.$$

for an ideal gas. In fact, γ is very easy to measure because the speed of sound in an ideal gas is written

$$c_s = \sqrt{\frac{\gamma p}{\rho}},$$

where ρ is the density. The extent of the agreement between γ calculated from Eq. and the experimental γ is quite remarkable.

Table. Specific Heats of Common Gases in Joules/mole/deg. (at 15 C and 1 atm.) From Reif.3

Gas	*Symbol*	c_V	γ	γ
		(experiment)	*(experiment)*	*(theory)*
Helium	He	12.5	1.666	1.666
Argon	Ar	12.5	1.666	1.666
Nitrogen	N_2	20.6	1.405	1.407
Oxygen	O_2	21.1	1.396	1.397
Carbon Dioxide	CO_2	28.2	1.302	1.298
Ethane		39.3	1.22	1.214

CALCULATION OF SPECIFIC HEATS

Now that we know the relationship between the specific heats at constant volume and constant pressure for an ideal gas, it would be nice if we could calculate either one of these quantities from first principles. Classical thermodynamics cannot help us here. However, it is quite easy to calculate the specific heat at constant volume using our knowledge of statistical physics. Recall, that the variation of the number of accessible states of an ideal gas with energy and volume is written

$$\Omega(E, V) \propto V^N \chi(E).$$

For the specific case of a monatomic ideal gas, we worked out a more exact expression for Ω,

$$\Omega(E, V) = B\, V^N E^{3N/2},$$

where B is a constant independent of the energy and volume. It follows that

$$\ln \Omega = \ln B + N \ln V + \frac{3N}{2} \ln E.$$

The temperature is given by

$$\frac{1}{kT} = \frac{\partial \ln \Omega}{\partial E} = \frac{3N}{2} \frac{1}{E},$$

so

$$E = \frac{3}{2} N\, kT.$$

Since, $N = v\, N_A$, and $N_A\, k = R$, we can rewrite the above expression as

$$E = \frac{3}{2} v\, RT,$$

where R = 8.3143 joules/mole/deg, is the ideal gas constant. The above formula tells us exactly how the internal energy of a monatomic ideal gas depends on its temperature.

The molar specific heat at constant volume of a monatomic ideal gas is clearly

$$c_V = \frac{1}{V} \left(\frac{\partial E}{\partial T} \right)_v = \frac{3}{2} R.$$

This has the numerical value

c_V = 12.47 joules/mole/deg.

Furthermore, we have

$$cp = cV + R = \frac{5}{2} R,$$

and

$$\gamma \equiv \frac{c_p}{c_V} = \frac{5}{3} = 1.667.$$

The predictions are borne out pretty well for the monatomic gases Helium and Argon. Note that the specific heats of polyatomic gases are larger than those of monatomic gases. This is because polyatomic molecules can rotate around their centres of mass, as well as translate, so polyatomic gases can store energy in the rotational, as well as the translational, energy states of their constituent particles.

ISOTHERMAL AND ADIABATIC EXPANSION

Suppose that the temperature of an ideal gas is held constant by keeping the gas in thermal contact with a heat reservoir. If the gas is allowed to expand quasi-statically under these so called isothermal conditions then the ideal equation of state tells us that

$$pV = \text{constant}.$$

This is usually called the isothermal gas law.

Suppose, now, that the gas is thermally isolated from its surroundings. If the gas is allowed to expand quasi-statically under these so called adiabatic conditions then it does work on its environment, and, hence, its internal energy is reduced, and its temperature changes. Let us work out the relationship between the pressure and volume of the gas during adiabatic expansion.

According to the first law of thermodynamics,

$$\bar{d}Q = v\, c_V\, dT + p dV = 0,$$

in an adiabatic process (in which no heat is absorbed). The ideal gas equation of state can be differentiated, yielding

$$p dV + V\, dp = v\, R dT.$$

The temperature increment dT can be eliminated between the above two expressions to give

$$0 = \frac{c_V}{R}(p dV + V\, dp) + p dV = \left(\frac{c_V}{R} + 1\right) p dV + \frac{c_V}{R} V\, dp,$$

which reduces to

$$(c_V + R) p\, dV + c_V\, V\, dp = 0$$

Dividing through by $c_V\, p\, V$ yields

$$\gamma \frac{dV}{V} + \frac{dp}{p} = 0,$$

where

$$\gamma \equiv \frac{c_p}{c_V} + \frac{c_V + R}{c_V}.$$

It turns out that c_V is a very slowly varying function of temperature in most gases. So, it is always a fairly good approximation to treat the ratio of specific heats γ as a constant, at least over a limited temperature range. If γ is constant then we can integrate Eq. to give

$$\gamma \ln V + \ln p = \text{constant},$$

or

$$pV\gamma = \text{constant}$$

This is the famous adiabatic gas law. It is very easy to obtain similar relationships between V and T and P and T during adiabatic expansion or contraction. Since $p = vRT/V$, the above formula also implies that

$$TV^{-1} = \text{constant},$$

and

$$p^{1-\gamma} T^{\gamma} = \text{constant}$$

Equations are all completely equivalent.

HYDROSTATIC EQUILIBRIUM OF THE TMOSPHERE

The gas which we are most familiar with in everyday life is, of course, the Earth's atmosphere. In fact, we can use the isothermal and adiabatic gas laws to explain most of the observable features of the atmosphere.

Let us, first of all, consider the hydrostatic equilibrium of the atmosphere. Consider a thin vertical slice of the atmosphere of cross-sectional area A which starts at height zabove ground level and extends to height $z + dz$. The upwards force exerted on this slice from the gas below is $p(z)A$, where $p(z)$ is the pressure at height z. Likewise, the downward force exerted by the gas above the slice is $p(z + dz) A$. The net upward force is clearly $[p(z) - p(z + dz)]A$. In equilibrium, this upward force must be balanced by the downward force due to the weight of the slice: this is $\rho\ Adz\ g$, where ρ is the density of the gas, and g is the acceleration due to gravity. In follows that the force balance condition can be written

$$[p(z) - p(z + dz)]A = \rho\ Adz\ g,$$

which reduces to

$$\frac{dp}{dz} = -\rho g.$$

This is called the equation of hydrostatic equilibrium for the atmosphere.

We can write the density of a gas in the following form,

$$\rho = \frac{v\,\mu}{V},$$

where μ is the molecular weight of the gas, and is equal to the mass of one mole of gas particles. For instance, the molecular weight of Nitrogen gas is 28 grams. The above formula for the density of a gas combined with the ideal gas law $pV = vRT$ yields

$$\rho = \frac{p\,\mu}{RT}.$$

It follows that the equation of hydrostatic equilibrium can be rewritten

$$\frac{dp}{p} = -\frac{\mu g}{RT}dz.$$

THE ISOTHERMAL ATMOSPHERE

As a first approximation, let us assume that the temperature of the atmosphere is uniform. In such an isothermal atmosphere, we can directly integrate the previous equation to give

$$p = p_0 \exp\left(-\frac{z}{z_0}\right).$$

Here, p_0 is the pressure at ground level ($z = 0$), which is generally about 1 bar, or 1 atmosphere ($10^5\ Nm^{-2}$ in Sl units). The quantity

$$z_0 = \frac{RT}{\mu g}$$

is called the isothermal scale-height of the atmosphere. At ground level, the temperature is on average about 15° centigrade, which is 288° kelvin on the absolute scale. The mean molecular weight of air at sea level is 29 (i.e., the molecular weight of a gas made up of 78% Nitrogen, 21%

Oxygen, and 1% Argon). The acceleration due to gravity is $9.81\ \mathrm{m\ s^{-2}}$ at ground level. Also, the ideal gas constant is 8.314 joules/mole/degree. Putting all of this information together, the isothermal scale-height of the atmosphere comes out to be about 8.4 kilometers.

We have discovered that in an isothermal atmosphere the pressure decreases exponentially with increasing height. Since the temperature is assumed to be constant, and $\rho \propto p/T$ it follows that the density also decreases exponentially with the same scale-height as the pressure.

According to Eq. the pressure, or density, decreases by a factor 10 every ln 10 z_0, or 19.3 kilometers, we move vertically upwards. Clearly, the effective height of the atmosphere is pretty small compared to the Earth's radius, which is about 6,400 kilometers. In other words, the atmosphere constitutes a very thin layer covering the surface of the Earth. Incidentally, this justifies our neglect of the decrease of g with increasing altitude.

One of the highest points in the United States of America is the peak of Mount Elbert in Colorado. This peak lies 14,432 feet, or about 4.4 kilometers, above sea level. At this altitude, our formula says that the air pressure should be about 0.6 atmospheres. Surprisingly enough, after a few days acclimatization, people can survive quite comfortably at this sort of pressure. In the highest inhabited regions of the Andes and Tibet, the air pressure falls to about 0.5 atmospheres. Humans can just about survive at such pressures.

However, people cannot survive for any extended period in air pressures below half an atmosphere. This sets an upper limit on the altitude of permanent human habitation, which is about 19,000 feet, or 5.8 kilometers, above sea level. Incidentally, this is also the maximum altitude at which a pilot can fly an unpressurized aircraft without requiring additional Oxygen.

The highest point in the world is, of course, the peak of Mount Everest in Nepal. This peak lies at an altitude of 29,028 feet, or 8.85 kilometers, above sea level, where we expect the air pressure to be a mere 0.35 atmospheres. This explains why

Mount Everest was only conquered after lightweight portable oxygen cylinders were invented. Admittedly, some climbers have subsequently ascended Mount Everest without the aid of additional oxygen, but this is a very foolhardy venture, because above 19,000 feet the climbers are slowly dying.

Commercial airliners fly at a cruising altitude of 32,000 feet. At this altitude, we expect the air pressure to be only 0.3 atmospheres, which explains why airline cabins are pressurized. In fact, the cabins are only pressurized to 0.85 atmospheres (which accounts for the "popping" of passangers ears during air travel). The reason for this partial pressurization is quite simple. At 32,000 feet, the pressure difference between the air in the cabin and that outside is about half an atmosphere.

Clearly, the walls of the cabin must be strong enough to support this pressure difference, which means that they must be of a certain thickness, and, hence, the aircraft must be of a certain weight. If the cabin were fully pressurized then the pressure difference at cruising altitude would increase by about 30%, which means that the cabin walls would have to be much thicker, and, hence, the aircraft would have to be substantially heavier. So, a fully pressurized aircraft would be more comfortable to fly in (because your ears would not "pop"), but it would also be far less economical to operate.

THE ADIABATIC ATMOSPHERE

Of course, we know that the atmosphere is not isothermal. In fact, air temperature falls quite noticeably with increasing altitude. In ski resorts, you are told to expect the temperature to drop by about 1 degree per 100 meters you go upwards. Many people cannot understand why the atmosphere gets colder the higher up you go.

They reason that as higher altitudes are closer to the Sun they ought to be hotter. In fact, the explanation is quite simple. It depends on three important properties of air. The first important property is that air is transparent to most, but by no means all, of the electromagnetic spectrum. In particular, most infrared radiation, which carries heat energy, passes straight through the lower atmosphere and heats the ground. In other words, the lower atmosphere is heated from below, not from above. The second important property of air is that it is constantly in motion. In fact, the lower 20 kilometers of the atmosphere (the so called troposphere) are fairly thoroughly mixed. You might think that this would imply that the atmosphere is isothermal. However, this is not the case because of the final important properly of air: i.e., it is a very poor conductor of heat. This, of course, is why woolly sweaters work: they trap a layer of air close to the body, and because air is such a poor conductor of heat you stay warm.

Imagine a packet of air which is being swirled around in the atmosphere. We would expect it to always remain at the same pressure as its surroundings, otherwise it would be mechanically unstable. It is also plausible that the packet moves around too quickly to effectively exchange heat with its surroundings, since air is very a poor heat conductor, and heat flow is consequently quite a slow process.

So, to a first approximation, the air in the packet is adiabatic. In a steady-state atmosphere, we expect that as the packet moves upwards, expands due to the reduced pressure, and cools adiabatically, its temperature always remains the same as that of its immediate surroundings. This means that we can use the adiabatic gas law to characterize the cooling of the atmosphere with increasing altitude. In this particular case, the most useful manifestation of the adiabatic law is

$$p^{1-\gamma}T^{\gamma} = \text{constant},$$

giving

$$\frac{dp}{p} = \frac{\gamma}{\gamma-1}\frac{dT}{T}.$$

Combining the above relation with the equation of hydrostatic equilibrium, we obtain

$$\frac{\gamma}{\gamma-1}\frac{dT}{T} = -\frac{\mu g}{RT}dz,$$

or

$$\frac{dT}{dz} = -\frac{\gamma-1}{\gamma}\frac{\mu g}{R}.$$

Now, the ratio of specific heats for air (which is effectively a diatomic gas) is about 1.4. Hence, we can calculate, from the above expression, that the temperature of the atmosphere decreases with increasing height at a constant rate of 9.8°centigrade per kilometer.

This value is called the adiabatic lapse rate of the atmosphere. Our calculation accords well with the "1 degree colder per 100 meters higher" rule of thumb used in ski resorts. The basic reason why air is colder at higher altitudes is that it expands as its pressure decreases with height. It, therefore, does work on its environment, without absorbing any heat (because of its low thermal conductivity), so its internal energy, and, hence, its temperature decreases.

According to the adiabatic lapse rate calculated above, the air temperature at the cruising altitude of airliners (32,000 feet) should be about –80° centigrade (assuming a sea level temperature of 15° centigrade). In fact, this is somewhat of an underestimate.

A more realistic value is about –60°centigrade. The explanation for this discrepancy is the presence of water vapour in the atmosphere. As air rises, expands, and cools, water vapour condenses out releasing latent heat which prevents the temperature from falling as rapidly with height as the adiabatic lapse rate would indicate.

In fact, in the Tropics, where the humidity is very high, the lapse rate of the atmosphere (i.e., the rate of decrease of temperature with altitude) is significantly less than the adiabatic value. The adiabatic lapse rate is only observed when the humidity is low. This is the case in deserts, in the Arctic (where water vapour is frozen out of the atmosphere), and, of course, in ski resorts.

Suppose that the lapse rate of the atmosphere differs from the adiabatic value. Let us ignore the complication of water vapour and assume that the atmosphere is dry. Consider a packet of air which moves slightly upwards from its equilibrium height.

The temperature of the packet will decrease with altitude according to the adiabatic lapse rate, because its expansion is adiabatic. We assume that the packet always maintains pressure balance with its surroundings. It follows that since $\rho T \propto p$, according to the ideal gas law, then

$$(\rho T)_{packet} = (\rho T)_{atmospher}.$$

If the atmospheric lapse rate is less than the adiabatic value then $T_{atmospher} > T_{pack}$ implying that $\rho_{packet} > \rho_{atmospher}$. So, the packet will be denser than its immediate surroundings, and will, therefore, tend to fall back to its original height. Clearly, an atmosphere whose lapse rate is less than the adiabatic value is stable. On the other hand, if the atmospheric lapse rate exceeds the adiabatic value then, after rising a little way, the packet will be less dense than its immediate surroundings, and will, therefore, continue to rise due to buoyancy effects.

Clearly, an atmosphere whose lapse rate is greater than the adiabatic value is unstable. This effect is of great importance in Meteorology. The normal stable state of the atmosphere is for the lapse rate to be slightly less than the adiabatic value. Occasionally, however, the lapse rate exceeds the adiabatic value, and this is always associated with extremely disturbed weather patterns.

Let us consider the temperature, pressure, and density profiles in an adiabatic atmosphere. We can directly integrate Eq. to give

$$T = T_0\left(1 - \frac{\gamma - 1}{\gamma}\frac{z}{z_0}\right),$$

where T_0 is the ground level temperature, and

$$z_0 = \frac{RT_0}{\mu g}$$

is the isothermal scale-height calculated using this temperature. The pressure profile is easily calculated from the adiabatic gas law $p^{1-\gamma} T^{\gamma}$ = constant, or $p \propto T^{\gamma/(\gamma-1)}$. It follows that

$$p = p_0\left(1 - \frac{\gamma - 1}{\gamma}\frac{z}{z_0}\right)^{\gamma/(\gamma-1)}.$$

Consider the limit $\gamma \to 1$. In this limit, Eq. yields T independent of height (i.e., the atmosphere becomes isothermal). We can evaluate Eq. in the limit as $\gamma \to 1$ using the mathematical identity

$$\mathrm{lt}_{m \to 0}(1 + mx)^{1/m} \equiv \exp(x).$$

We obtain

$$p = p_0 \exp\left(-\frac{z}{z_0}\right),$$

which, not surprisingly, is the predicted pressure variation in an isothermal atmosphere. In reality, the ratio of specific heats of the atmosphere is not unity, it is about 1.4 (i.e., the ratio for diatomic gases), which implies that in the real atmosphere

$$p = p_0 \left(1 - \frac{z}{3.5\, z_0}\right)^{3.5}.$$

In fact, this formula gives very similar results to the exponential formula, Eq. for heights below one scale-height (i.e., $z < z_0$). For heights above one scale-height, the exponential formula tends to predict too low a pressure. So, in an adiabatic atmosphere, the pressure falls off less quickly with altitude than in an isothermal atmosphere, but this effect is only really noticeable at pressures significantly below one atmosphere. In fact, the isothermal formula is a pretty good approximation below altitudes of about 10 kilometers. Since $\rho \propto p/T$, the variation of density with height goes like

$$p = p_0 \left(1 - \frac{\gamma - 1}{\gamma} \frac{z}{z_0}\right)^{1/(\gamma-1)}$$

where ρ_0 is the density at ground level. Thus, the density falls off more rapidly with altitude than the temperature, but less rapidly than the pressure.

Note that an adiabatic atmosphere has a sharp upper boundary. Above height $z_1 = [\gamma/(\gamma - 1]\, z_0$ the temperature, pressure, and density are all zero: i.e., there is no atmosphere. For real air, with $\gamma = 1.4$, $z_1 \simeq 3.5\, z_0 \simeq 29.4$ kilometers. This behaviour is quite different to that of an isothermal atmosphere, which has a diffuse upper boundary. In reality, there is no sharp upper boundary to the atmosphere. The adiabatic gas law does not apply above about 20 kilometers (i.e., in the stratosphere) because at these altitudes the air is no longer strongly mixed. Thus, in the stratosphere the pressure falls off exponentially with increasing height.

In conclusion, we have demonstrated that the temperature of the lower atmosphere should fall off approximately linearly with increasing height above ground level, whilst the pressure should fall off far more rapidly than this, and the density should fall off at some intermediate rate. We have also shown that the lapse rate of the temperature should be about 10°centigrade per kilometer in dry air, but somewhat less than this in wet air. In fact, all off these predictions are, more or less, correct. It is amazing that such accurate predictions can be obtained from the two simple laws, pV = constant for an isothermal gas, and pV^{γ} = constant for an adiabatic gas.

Heat Engines

Thermodynamics was invented, almost by accident, in 1825 by a young French engineer called Sadi Carnot who was investigating the theoretical limitations on the efficiency of steam engines. Although we are not particularly interested in steam engines, nowadays, it is still highly instructive to review some of Carnot's arguments.

We know, by observation, that it is possible to do mechanical work w upon a device M, and then to extract an equivalent amount of heat q, which goes to increase the internal energy of some heat reservoir. (Here, we use small letters w and q to denote intrinsically positive amounts of work and heat, respectively.)

An example of this is Joule's classic experiment by which he verified the first law of thermodynamics: a paddle wheel is spun in a liquid by a falling weight, and the work done by the weight on the wheel is converted into heat, and absorbed by the liquid.

Carnot's question was this: is it possible to reverse this process and build a device, called a heat engine, which extracts heat energy from a reservoir and converts it into useful macroscopic work? For instance, is it possible to extract heat from the ocean and use it to run an electric generator?

There are a few caveats to Carnot's question. First of all, the work should not be done at the expense of the heat engine itself, otherwise the conversion of heat into work could not continue indefinitely. We can ensure that this is the case if the heat engine performs some sort of cycle, by which it periodically returns to the same macrostate, but, in the meantime, has extracted heat from the reservoir and done an equivalent amount of useful work. Furthermore, a cyclic process seems reasonable because we know that both steam engines and internal combustion engines perform continuous cycles.

The second caveat is that the work done by the heat engine should be such as to change a single parameter of some external device (e.g., by lifting a weight) without doing it at the expense of affecting the other degrees of freedom, or the entropy, of that device. For instance, if we are extracting heat from the ocean to generate electricity, we want to spin the shaft of the electrical generator without increasing the generator's entropy; i.e., causing the generator to heat up or fall to bits.

Let us examine the feasibility of a heat engine using the laws of thermodynamics. Suppose that a heat engine M performs a single cycle. Since M has returned to its initial macrostate, its internal energy is unchanged, and the first law of thermodynamics tell us that the work done by the engine w must equal the heat extracted from the reservoir q, so

$$w = q.$$

The above condition is certainly a necessary condition for a feasible heat engine, but is it also a sufficient condition? In other words, does every device which satisfies this condition actually work? Let us think a little more carefully about what we are actually expecting a heat engine to do. We want to construct a device which will extract energy from a heat reservoir, where it is randomly distributed over very many degrees of freedom, and convert it into energy distributed over a single degree of freedom associated with some parameter of an external device.

Once we have expressed the problem in these terms, it is fairly obvious that what we are really asking for is a spontaneous transition from a probable to an improbable state, which we know is forbidden by the second law of thermodynamics.

So, unfortunately, we cannot run an electric generator off heat extracted from the ocean, because it is like asking all of the molecules in the ocean, which are jiggling about every which way, to all suddenly jig in the same direction, so as to exert a force on some lever, say, which can then be converted into a torque on the generator shaft.

We know from our investigation of statistical thermodynamics that such a process is possible, in principle, but is fantastically improbable.

The improbability of the scenario just outlined is summed up in the second law of thermodynamics. This says that the total entropy of an isolated system can never spontaneously decrease, so

$$\Delta S \geq 0.$$

For the case of a heat engine, the isolated system consists of the engine, the reservoir from which it extracts heat, and the outside device upon which it does work. The engine itself returns periodically to the same state, so its entropy is clearly unchanged after each cycle.

We have already specified that there is no change in the entropy of the external device upon which the work is done. On the other hand, the entropy change per cycle of the heat reservoir, which is at absolute temperature T_1, say, is given by

$$\Delta S_{\text{reservoir}} = \oint \frac{đ}{T_1} = -\frac{q}{T_1},$$

where $đQ$ is the infinitesimal heat absorbed by the reservoir, and the integral is taken over a whole cycle of the heat engine. The integral can be converted into the expression $-q/T_1$ because the amount of heat extracted by the engine is assumed to be too small to modify the temperature of the reservoir (this is the definition of a heat reservoir), so that T_1 is a constant during the cycle. The second law of thermodynamics clearly reduces to

$$\frac{-q}{T_1} \geq 0$$

or, making use of the first law of thermodynamics,

$$\frac{q}{T_1} = \frac{w}{T_1} \leq 0.$$

Since we wish the work w done by the engine to be positive, the above relation clearly cannot be satisfied, which proves that an engine which converts heat directly into work is thermodynamically impossible.

A perpetual motion device, which continuously executes a cycle without extracting heat from, or doing work on, its surroundings, is just about possible according to Eq. In fact, such a device corresponds to the equality sign in Eq. which means that it must be completely reversible. In reality, there is no such thing as a completely reversible engine. All engines, even the most efficient, have frictional losses which make them, at least, slightly irreversible.

Thus, the equality sign in Eq. corresponds to an asymptotic limit which reality can closely approach, but never quite attain. It follows that a perpetual motion device is thermodynamically impossible. Nevertheless, the U.S. patent office receives about 100 patent applications a year regarding perpetual motion devices. The British patent office, being slightly less open-minded that its American counterpart, refuses to entertain such applications on the basis that perpetual motion devices are forbidden by the second law of thermodynamics.

According to Eq. there is no thermodynamic objection to a heat engine which runs backwards, and converts work directly into heat. This is not surprising, since we know that this is essentially what frictional forces do. Clearly, we have, here, another example of a natural process which is fundamentally irreversible according to the second law of thermodynamics. In fact, the statement it is impossible to construct a perfect heat engine which converts heat directly into work is called Kelvin's formulation of the second law.

We have demonstrated that a perfect heat engine, which converts heat directly into work, is impossible.

But, there must be some way of obtaining useful work from heat energy, otherwise steam engines would not operate. Well, the reason that our previous scheme did not work was that it decreased the entropy of a heat reservoir, at some temperature T_1, by extracting an amount of heat q per cycle, without any compensating increase in the entropy of anything else, so the second law of thermodynamics was violated. How can we remedy this situation?

We still want the heat engine itself to perform periodic cycles (so, by definition, its entropy cannot increase over a cycle), and we also do not want to increase the entropy of the external device upon which the work is done. Our only other option is to increase the entropy of some other body. In Carnot's analysis, this other body is a second heat reservoir at temperature T_2.

We can increase the entropy of the second reservoir by dumping some of the heat we extracted from the first reservoir into it.

Suppose that the heat per cycle we extract from the first reservoir is q_1, and the heat per cycle we reject into the second reservoir is q_2. Let the work done on the external device be w per cycle. The first law of thermodynamics tells us that

$$q_1 = w + q_2.$$

Note that $q_2 < q_1$ if positive (i.e., useful) work is done on the external device. The total entropy change per cycle is due to the heat extracted from the first reservoir and the heat dumped into the second, and has to be positive (or zero) according to the second law of thermodynamics. So,

$$\Delta S = \frac{-q_1}{T_1} + \frac{q_2}{T_2} \geq 0.$$

We can combine the previous two equations to give

$$\frac{-q_1}{T_1} + \frac{q_2 - w}{T_2} \geq 0,$$

or

$$\frac{w}{T_2} \leq q_1\left(\frac{1}{T_2} - \frac{1}{T_2}\right).$$

It is clear that the engine is only going to perform useful work (i.e., w is only going to be positive) if $T_2 < T_1$. So, the second reservoir has to be colder than the first if the heat dumped into the former is to increase the entropy of the Universe more than the heat extracted from the latter decreases it. It is useful to define the efficiency η of a heat engine. This is the ratio of the work done per cycle on the external device to the heat energy absorbed per cycle from the first reservoir. The efficiency of a perfect heat engine is unity, but we have already shown that such an engine is impossible. What is the efficiency of a realizable engine? It is clear from the previous equation that

$$\eta \equiv \frac{w}{q_2} \leq 1 \frac{T_2}{T_1} = \frac{T_1 - T_2}{T_1}.$$

Note that the efficiency is always less than unity. A real engine must always reject some energy into the second heat reservoir in order to satisfy the second law of thermodynamics, so less energy is available to do external work, and the efficiency of the engine is reduced. The equality sign in the above expression corresponds to a completely reversible heat engine (i.e., one which is quasi-static). It is clear that real engines, which are always irreversible to some extent, are less efficient than reversible engines. Furthermore, all reversible engines which operate between the two temperatures T_1 and T_2 must have the same efficiency,

$$\eta = \frac{T_1 - T_2}{T_1},$$

irrespective of the way in which they operate.

Let us consider how we might construct one of these reversible heat engines. Suppose that we have some gas in a cylinder equipped with a frictionless piston. The gas is not necessarily a perfect gas. Suppose that we also have two heat reservoirs at temperatures T_1 and T_2 (where $T_1 > T_2$). These reservoirs might take the form of large water baths. Let us start off with the gas in thermal contact with the first reservoir. We now pull the piston out very slowly so that heat energy flows reversibly into the gas from the reservoir.

Let us now thermally isolate the gas and slowly pull out the piston some more. During this adiabatic process the temperature of the gas falls (since there is no longer any heat flowing into it to compensate for the work it does on the piston). Let us continue this process until the temperature of the gas falls to T_2.

We now place the gas in thermal contact with the second reservoir and slowly push the piston in. During this isothermal process heat flows out of the gas into the reservoir. We next thermally isolate the gas a second time and slowly compress it some more. In this process the temperature of the gas increases.

We stop the compression when the temperature reaches T_1. If we carry out each step properly we can return the gas to its initial state and then repeat the cycle ad infinitum. We now have a set of reversible processes by which a quantity of heat is extracted from the first reservoir and a quantity of heat is dumped into the second.

We can best evaluate the work done by the system during each cycle by plotting out the locus of the gas in a p–V diagram. The locus takes the form of a closed curve.

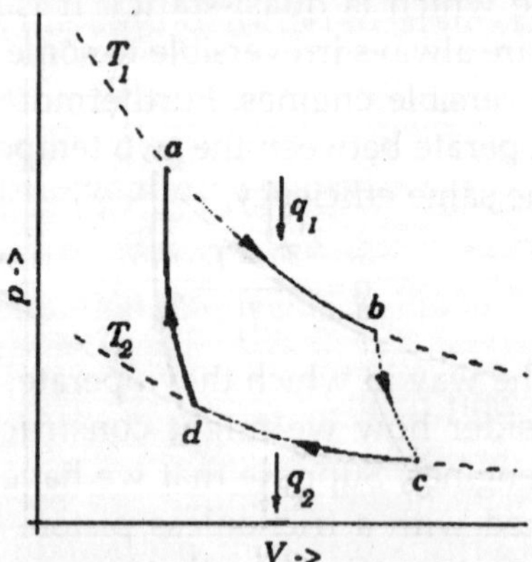

Fig. An ideal Gas Carnot Engine.

The net work done per cycle is the "area" contained inside

this curve, since $đW = p\,dV$ [if p is plotted vertically and V horizontally, then $p\,dV$ is clearly an element of area under the curve $p\,(V)$]. The engine we have just described is called a Carnot engine, and is the simplest conceivable device capable of converting heat energy into useful work.

For the specific case of an ideal gas, we can actually calculate the work done per cycle, and, thereby, verify Eq. Consider the isothermal expansion phase of the gas. For an ideal gas, the internal energy is a function of the temperature alone. The temperature does not change during isothermal expansion, so the internal energy remains constant, and the net heat absorbed by the gas must equal the work it does on the piston. Thus,

$$q_1 = \int_a^b p dV,$$

where the expansion takes the gas from state a to state b. Since $pV = vRT$, for an ideal gas, we have

$$q_1 = \int_a^b v\,RT_1 \frac{dV}{V} = vRT_1 \ln \frac{V_b}{V_a}.$$

Likewise, during the isothermal compression phase, in which the gas goes from state c to state d, the net heat rejected to the second reservoir is

$$q_1 = vRT_2 \ln \frac{V_c}{V_d}.$$

Now, during adiabatic expansion or compression

$$TV^{\gamma-1} = \text{constant}.$$

It follows that during the adiabatic expansion phase, which takes the gas from state b to state c,

$$T_1 V_b^{\gamma-1} = T_2 V_c^{\gamma-1}.$$

Likewise, during the adiabatic compression phase, which takes the gas from state d to state a,

$$T_1 V_a^{\gamma-1} = T_2 V_d^{\gamma-1}.$$

If we take the ratio of the previous two equations we obtain

$$\frac{V_b}{V_a} = \frac{V_c}{V_d}.$$

Hence, the work done by the engine, which we can calculate using the first law of thermodynamics,

$$w = q_1 - q_2,$$

is

$$w = vR(T_1 - T_2)\ln\frac{V_b}{V_a}.$$

Thus, the efficiency of the engine is

$$\eta = \frac{w}{q_1} = \frac{T_1 - T_2}{T_1}$$

which, not surprisingly, is exactly the same as Eq.

The engine described above is very idealized. Of course, real engines are far more complicated than this. Nevertheless, the maximum efficiency of an ideal heat engine places severe constraints on real engines. Conventional power stations have many different "front ends" (e.g., coal fired furnaces, oil fired furnaces, nuclear reactors), but their "back ends" are all very similar, and consist of a steam driven turbine connected to an electric generator.

The "front end" heats water extracted from a local river and turns it into steam, which is then used to drive the turbine, and, hence, to generate electricity.

Finally, the steam is sent through a heat exchanger so that it can heat up the incoming river water, which means that the incoming water does not have to be heated so much by the "front end."

At this stage, some heat is rejected to the environment, usually as clouds of steam escaping from the top of cooling towers. We can see that a power station possesses many of the same features as our idealized heat engine.

There is a cycle which operates between two temperatures. The upper temperature is the temperature to which the steam is heated by the "front end," and the lower temperature is the temperature of the environment into which heat is rejected. Suppose that the steam is only heated to 100°C (or 373°K), and the temperature of the environment is 15°C (or 288°K). It follows from Eq. that the maximum possible efficiency of the steam cycle is

$$\eta = \frac{373-288}{373} \simeq 0.23.$$

So, at least 77% of the heat energy generated by the "front end" goes straight up the cooling towers! Not be surprisingly, commercial power stations do not operate with 100°C steam. The only way in which the thermodynamic efficiency of the steam cycle can be raised to an acceptable level is to use very hot steam (clearly, we cannot refrigerate the environment). Using 400°C steam, which is not uncommon, the maximum efficiency becomes

$$\eta = \frac{673-288}{673} \simeq 0.57,$$

which is more reasonable. In fact, the steam cycles of modern power stations are so well designed that they come surprisingly close to their maximum thermodynamic efficiencies.

REFRIGERATORS

Let us now move on to consider refrigerators. An idealized refrigerator is an engine which extracts heat from a cold heat reservoir (temperature T_2, say) and rejects it to a somewhat hotter heat reservoir, which is usually the environment (temperature T_1, say). To make this machine work we always have to do some external work on the engine. For instance, the refrigerator in your home contains a small electric pump which does work on the freon (or whatever) in the cooling circuit.

We can see that, in fact, a refrigerator is just a heat engine run in reverse. Hence, we can immediately carry over most of our heat engine analysis.

Let q_2 be the heat absorbed per cycle from the colder reservoir, q_1 the heat rejected per cycle into the hotter reservoir, and w the external work done per cycle on the engine. The first law of thermodynamics tells us that

$$w + q_2 = q_1.$$

The second law says that

$$\frac{q_1}{T_1} + \frac{-q_2}{T_2} \geq 0.$$

We can combine these two laws to give

$$\frac{w}{T_1} \geq q_2\left(\frac{1}{T_2} - \frac{1}{T_1}\right).$$

The most sensible way of defining the efficiency of a refrigerator is as the ratio of the heat extracted per cycle from the cold reservoir to the work done per cycle on the engine. With this definition

$$\eta = \frac{T_2}{T_1 - T_2}.$$

We can see that this efficiency is, in general, greater than unity. In other words, for one joule of work done on the engine, or pump, more than one joule of energy is extracted from whatever it is we are cooling.

Clearly, refrigerators are intrinsically very efficient devices. Domestic refrigerators cool stuff down to about 4° C(277° K) and reject heat to the environment at about 15°C (288° K). The maximum theoretical efficiency of such devices is

$$\eta = \frac{277}{288 - 277} = 25.2.$$

So, for every joule of electricity we put into a refrigerator we can extract up to 25 joules of heat from its contents.

Chapter 6

Laws of Thermodynamics

0TH LAW OF THERMODYNAMICS

Thermal Equilibrium: A system with many microscopic components (for example, a gas, a liquid, a solid with many molecules) that is isolated from all forms of energy exchange and left alone for a "long time" moves toward a state of thermal equilibrium. A system in thermal equilibrium is characterized by a set of macroscopic quantities that depend on the system in question and characterize its "state" (such as pressure, volume, density) that do not change in time.

Two systems are said to be in (mutual) thermal equilibrium if, when they are placed in "thermal contact" (basically, contact that permits the exchange of energy between them), their state variables do not change. A system is said to be in thermal equilibrium when its temperature is stable (i.e does not change over time). Mathematically, this could be expressed as:

$$\frac{dT}{dt} = 0.$$

Zeroth Law of Thermodynamics: If system A is in thermal equilibrium with system C, and system B is in thermal equilibrium with system C, then system A is in thermal equilibrium with system B.

Temperature and Thermometers: The point of the Zeroth Law is that it is the basis of the thermometer. A thermometer is a portable device whose thermal state is related linearly to some simple property, for example its density or pressure. Once a suitable temperature scale is defined for the device, one can use it to measure the temperature of a variety of disparate systems in thermal equilibrium. Temperature thus characterizes thermal equilibrium.

Temperature Scales:

- *Fahrenheit:* This is one of the oldest scales, and is based on the coldest temperature that could be achieved with a mix of ice and alcohol. In it the freezing point of water is at 32° F, the boiling point of water is at 212° F.
- *Celsius or Centigrade:* This is a very sane system, where the freezing point of water is at 0° C and the boiling point is at 100° C. The degree size is thus as big 9/5 as the Fahrenheit degree.
- *Kelvin or Absolute:* 0°K is the lowest possible temperature, where the internal energy of a system is at its absolute minimum. The degree size is the same as that of the Centigrade or Celsius scale. This makes the freezing point of water at atmospheric pressure 273.16°K, the boiling point at 373.16°K.
- Thermal Expansion

$$\Delta L = \alpha L \Delta T$$

- where α is the coefficient of linear expansion. If one applies this in three dimensions:

$$\Delta V = \beta V \Delta T$$

- where $\beta = 3\alpha$.
- Ideal Gas Law

$$PV = nRT = NkT$$

- where J/mol-K, and $k = R/N_A = 1.38 \times 10^{-23}$ J/K.

THERMAL EQUILIBRIUM BETWEEN MANY SYSTEMS

Many systems are said to be in equilibrium if the small exchanges (due to Brownian motion, for example) between them do not lead to a net change in the total energy summed over all systems.

Consider N systems in adiabatic isolation from the rest of the universe (i.e no heat exchange is possible outside of these N systems), all of which have a constant volume and composition, who can only exchange heat with one another. The combined first and second laws relate the fluctuations in total energy äU to the temperature of the ith system Ti and the entropy fluctuation in the ith system äSi by,

$$\delta U = \sum_i^N T_i \delta S_i .$$

The adiabatic isolation of the system from the remaining universe requires that the total sum of the entropy fluctuations vanishes,

$$\sum_i^N \delta S_i = 0 ,$$

that is, entropy can only be exchanged between the *N* systems. This constraint can be used to re-arrange the expression for the total energy fluctuation to give,

$$\delta U = \sum_i^N (T_i - T_j)\, \delta S_i ,$$

where Tj is the temperature of any system j we may choose to single out among the N systems. Finally, equilibrium requires the total fluctuation in energy to vanish, so we arrive at,

$$\sum_i^N (T_i - T_j)\, \delta S_i = 0$$

which can be thought of as the vanishing of the product of an anti-symmetric matrix Ti " Tj and a vector of entropy fluctuations äSi. In order for a non-trivial solution to exist,

$$\delta S_i \neq 0,$$

the determinant of the matrix formed by Ti " Tj must vanish for all choices of N. However, according to Jacobi's theorem, the determinant of a NxN anti-symmetric matrix is always zero if N is odd, although for N even we find that all of the entries must vanish, Ti " Tj = 0, in order to obtain a vanishing determinant, and hence Ti = Tj at equilibrium. This non-intuitive result means that an odd number of systems are always in equilibrium regardless of their temperatures and entropy fluctuations, while equality of temperatures is only required between an even number of systems to achieve equilibrium in the presence of entropy fluctuations.

The zeroth law solves this odd vs. even paradox, because it can readily be used to reduce an odd-numbered system to an even number by considering any three of the N systems and eliminating one by application of its principle, and hence reduce the problem to even N which subsequently leads to the same equilibrium condition that we expect in every case, i.e., Ti = Tj. The same result applies to fluctuations in any extensive quantity, such as volume (yielding the equal pressure condition), or fluctuations in mass (leading to equality of chemical potentials), and therefore the zeroth law carries implications for a great deal more than temperature alone. In general, the zeroth law breaks a certain kind of anti-symmetry that exists in the first and second laws.

Temperature and the Zeroth Law

In the space of thermodynamic parameters, zones of constant temperature will form a surface, which provides a natural order of nearby surfaces. It is then simple to construct a global temperature function that provides a continuous ordering of states. Note that the dimensionality of a surface of constant temperature is one less than the number of thermodynamic parameters (thus, for an ideal gas described with 3 thermodynamic parameter P, V and n, they are 2D surfaces). The temperature so defined may indeed not look like the Celsius temperature scale, but it is a temperature function.

For example, if two systems of ideal gas are in equilibrium, then P1V1/N1 = P2V2/N2 where Pi is the pressure in the ith system, Vi is the volume, and Ni is the 'amount' (in moles, or simply number of atoms) of gas.

The surface PV / N = const defines surfaces of equal temperature, and the obvious (but not only) way to label them is to define T so that PV / N = RT where R is some constant. These systems can now be used as a thermometer to calibrate other systems.

FIRST LAW OF THERMODYNAMICS

The first law of thermodynamics is the application of the conservation of energy principle to heat and thermodynamic processes:

The change in internal energy of a system is equal to the heat added to the system minus the work done by the system.

$$\Delta U = Q - W$$

Change in internal energy — Heat added to the system — Work done by the system

The first law makes use of the key concepts of internal energy, heat, and system work. It is used extensively in the discussion of heat engines.

It is typical for chemistry texts to write the first law as $\Delta U = Q + W$. It is the same law, of course - the thermodynamic expression of the conservation of energy principle. It is just that W is defined as the work done on the system instead of work done by the system. In the context of physics, the common scenario is one of adding heat to a volume of gas and using the expansion of that gas to do work, as in the pushing down of a piston in an internal combustion engine. In the context of chemical reactions and process, it may be more common to deal with situations where work is done on the system rather than by it.

ENTHALPY

Four quantities called "thermodynamic potentials" are useful in the chemical thermodynamics of reactions and non-cyclic processes. They are internal energy, the enthalpy, the Helmholtz free energy and the Gibbs free energy. Enthalpy is defined by

$$H = U + PV$$

where P and V are the pressure and volume, and U is internal energy. Enthalpy is then a precisely measurable state variable, since it is defined in terms of three other precisely definable state variables. It is somewhat parallel to the first law of thermodynamics for a constant pressure system

$$Q = \Delta U + P\Delta V \text{ since in this case } Q = \Delta H$$

It is a useful quantity for tracking chemical reactions. If as a result of an exothermic reaction some energy is released to a system, it has to show up in some measurable form in terms of the state variables. An increase in the enthalpy H = U + PV might be associated with an increase in internal energy which could be measured by calorimetry, or with work done by the system, or a combination of the two.

The internal energy U might be thought of as the energy required to create a system in the absence of changes in temperature or volume.

But if the process changes the volume, as in a chemical reaction which produces a gaseous product, then work must be done to produce the change in volume. For a constant pressure process the work you must do to produce a volume change ÄV is PÄV. Then the term PV can be interpreted as the work you must do to "create room" for the system if you presume it started at zero volume.

THERMODYNAMIC PROPERTIES OF SELECTED SUBSTANCES

For one Mole at 298K and 1 Atmosphere Pressure3

Substance (form)	*Enthalpy "fH (kJ)*	*Gibbs "fG (kJ)*	*Entropy (J/ K*	*Specific heat CP(J/K)*	*Volume V(cm3)*
Al (s)	0	0	28.33	24.35	9.99
Al2SiO5 (kyanite)	-2594.3	-2443.9	83.81	121.71	44.09
Al2SiO5 (andalusite)	-2590.3	-2442.7	93.22	122.72	51.53
Al2SiO5 (sillimanite)	-2587.8	-2441	96.11	124.52	49.9
Ar (g)	0	0	154.84	20.79	...
C (graphite)	0	0	5.74	8.53	5.3
C (diamond)	1.895	2.9	2.38	6.11	3.42
CH4 (g)	-74.81	-50.72	186.26	35.31	...
C2H6 (g)	-84.68	-32.82	229.6	52.63	...
C3H8 (g)	-103.85	-23.49	269.91	73.5	...
C2H5OH (l)	-277.69	-174.78	160.7	111.46	58.4
C6H12O6 (glucose)	-1268	-910	212	115	...
CO (g)	-110.53	-137.17	197.67	29.14	...

Substance (form)	*Enthalpy "fH (kJ)*	*Gibbs "fG (kJ)*	*Entropy (J/ K*	*Specific heat CP(J/K)*	*Volume V(cm3)*
CO2 (g)	-393.51	-394.36	213.74	37.11	...
H2CO3 (aq)	-699.65	-623.08	187.4	...	...
HCO3- (aq)	-691.99	-586.77	91.2	...	...
Ca2+ (aq)	-542.83	-553.58	-53.1	...	...
CaCO3 (calcite)	-1206.9	-1128.8	92.9	81.88	36.93
CaCO3 (aragonite)	-1207.1	-1127.8	88.7	81.25	34.15
CaCl2 (s)	-795.8	-748.1	104.6	72.59	51.6
Cl2 (g)	0	0	223.07	33.91	...
Cl- (aq)	-167.16	-131.23	56.5	-136.4	17.3
Cu (s)	0	0	33.15	24.44	7.12
Fe (s)	0	0	27.28	25.1	7.11
H2 (g)	0	0	130.68	28.82	...
H (g)	217.97	203.25	114.71	20.78	...
H+ (aq)	0	0	0	0	...
H2O (l)	-285.83	-237.13	69.91	75.29	18.068
H2O (g)	-241.82	-228.57	188.83	33.58	...
He (g)	0	0	126.15	20.79	...
Hg (l)	0	0	76.02	27.98	14.81
N2 (g)	0	0	191.61	29.12	...
NH3 (g)	-46.11	-16.45	192.45	35.06	...
Na+(aq)	-240.12	-261.91	59	46.4	-1.2
NaCl (s)	-411.15	-384.14	72.13	50.5	27.01
NaAlSi3O8 (albite)	-3935.1	-3711.5	207.4	205.1	100.07
NaAlSi2O6 (jadeite)	-3030.9	-2852.1	133.5	160	60.4
Ne (g)	0	0	146.33	20.79	...
O2 (g)	0	0	205.14	29.38	...
O2 (aq)	-11.7	16.4	110.9	...	...
OH- (aq)	-229.99	-157.24	-10.75	-148.5	...
Pb (s)	0	0	64.81	26.44	18.3
PbO2(s)	-277.4	-217.33	68.6	64.64	...
PbSO4(s)	-920	-813	148.5	103.2	...
SO42- (aq)	-909.27	-744.53	20.1	-293	...
HSO4- (aq)	-887.34	-755.91	131.8	-84	...
SiO2	-910.94	-856.64	41.84	44.43	22.69
H4SiO4(aq)	-1449.4	-1307.7	215.13	468.98	...

SYSTEM WORK

When work is done by a thermodynamic system, it is ususlly a gas that is doing the work. The work done by a gas at constant pressure is:

$$W = P\Delta V$$

The line from a to b represents an expansion of a gas at constant pressure. The work done is the area under the curve.

For non-constant pressure, the work can be visualized as the area under the pressure-volume curve which represents the process taking place. The more general expression for work done is:

The integral expression gives the exact area under the curve which is equal to the work.

$$W = \int_{V_1}^{V_2} P\,dV$$

Work done by a system decreases the internal energy of the system, as indicated in the First Law of Thermodynamics. System work is a major focus in the discussion of heat engines.

DESCRIPTION

The first law of thermodynamics basically states that a thermodynamic system can store or hold energy and that this internal energy is conserved.

Heat is a process by which energy is added to a system from a high-temperature source, or lost to a low-temperature sink. In addition, energy may be lost by the system when it does mechanical work on its surroundings, or conversely, it may gain energy as a result of work done on it by its surroundings.

The first law states that this energy is conserved: The change in the internal energy is equal to the amount added by heating minus the amount lost by doing work on the environment. The first law can be stated mathematically as:

$$dU = dQ - \delta W$$

where dU is a small increase in the internal energy of the system, δQ is a small amount of heat added to the system, and äW is a small amount of work done by the system.

The δ's before the heat and work terms are used to indicate that they describe an increment of energy which is to be interpreted somewhat differently than the dU increment of internal energy. Work and heat are processes which add or subtract energy, while the internal energy U is a particular form of energy associated with the system. Thus the term "heat energy" for δQ means "that amount of energy added as the result of heating" rather than referring to a particular form of energy.

Likewise, the term "work energy" for δw means "that amount of energy lost as the result of work". The most significant result of this distinction is the fact that one can clearly state the amount of internal energy possessed by a thermodynamic system, but one cannot tell how much energy has flowed into or out of the system as a result of its being heated or cooled, nor as the result of work being performed on or by the system.

The first explicit statement of the first law of thermodynamics was given by Rudolf Clausius in 1850: "There is a state function E, called 'energy', whose differential equals the work exchanged with the surroundings during an adiabatic process."

Note that the above formulation is favored by engineers and physicists. Chemists prefer a second form, in which the work term δw is defined as the work done on the system, and therefore insert a plus sign in the above equation before the work term. This article will use the first definition exclusively.

MATHEMATICAL FORMULATION

The mathematical statement of the first law is given by:

$$dU = dQ - \delta W$$

where dU is the infinitesimal increase in the internal energy of the system, δQ is the infinitesimal amount of heat added to the system, and δw is the infinitesimal amount of work done by the system. The infinitesimal heat and work are denoted by δ rather than d because, in mathematical terms, they are not exact differentials. In other words, they do not describe the state of any system.

The integral of an inexact differential depends upon the particular "path" taken through the space of thermodynamic parameters while the integral of an exact differential depends only upon the initial and final states. If the initial and final states are the same, then the integral of an inexact differential may or may not be zero, but the integral of an exact differential will always be zero. The path taken by a thermodynamic system through state space is known as a thermodynamic process.

An expression of the first law can be written in terms of exact differentials by realizing that the work that a system does is, in case of a reversible process, equal to its pressure times the infinitesimal change in its volume.

In other words δw = PdV where P is pressure and V is volume. Also, for a reversible process, the total amount of heat added to a system can be expressed as δQ = TdS where T is temperature and S is entropy. Therefore, for a reversible process, :

$$dU = TdS - PdV.$$

Since U, S and V are thermodynamic functions of state, the above relation holds also for non-reversible changes. The above equation is known as the fundamental thermodynamic relation.

In the case where the number of particles in the system is not necessarily constant and may be of different types, the first law is written:

$$dU = \delta Q - \delta W + \sum_i \mu_i dN_i$$

where dNi is the (small) number of type-i particles added to the system, and μi is the amount of energy added to the system when one type-i particle is added, where the energy of that particle is such that the volume and entropy of the system remains unchanged. μi is known as the chemical potential of the type-i particles in the system. The statement of the first law, using exact differentials is now:

$$dU = TdS - PdV + \sum_i \mu_i dN_i.$$

If the system has more external variables than just the

volume that can change, the fundamental thermodynamic relation generalizes to:

$$dU = TdS - \sum_i X_i dx_i + \sum_i \mu_i dN_j.$$

Here the Xi are the generalized forces corresponding to the external variables xi.

A useful idea from mechanics is that the energy gained by a particle is equal to the force applied to the particle multiplied by the displacement of the particle while that force is applied. Now consider the first law without the heating term: dU = – PdV. The pressure P can be viewed as a force (and in fact has units of force per unit area) while dV is the displacement (with units of distance times area). We may say, with respect to this work term, that a pressure difference forces a transfer of volume, and that the product of the two (work) is the amount of energy transferred as a result of the process.

It is useful to view the TdS term in the same light: With respect to this heat term, a temperature difference forces a transfer of entropy, and the product of the two (heat) is the amount of energy transferred as a result of the process. Here, the temperature is known as a "generalized" force (rather than an actual mechanical force) and the entropy is a generalized displacement.

Similarly, a difference in chemical potential between groups of particles in the system forces a trasfer of particles, and the corresponding product is the amount of energy transferred as a result of the process. For example, consider a system consisting of two phases: liquid water and water vapor. There is a generalized "force" of evaporation which drives water molecules out of the liquid. There is a generalized "force" of condensation which drives vapor molecules out of the vapor. Only when these two "forces" (or chemical potentials) are equal will there be equilibrium, and the net transfer will be zero.

The two thermodynamic parameters which form a generalized force-displacement pair are termed "conjugate variables". The two most familiar pairs are, of course, pressure-volume, and temperature-entropy.

Types of Thermodynamic Processes

Paths through the space of thermodynamic variables are often specified by holding certain thermodynamic variables constant. It is useful to group these processes into pairs, in which each variable held constant is one member of a conjugate pair.

The pressure-volume conjugate pair is concerned with the transfer of mechanical or dynamic energy as the result of work.

- An isobaric process occurs at constant pressure. An example would be to have a movable piston in a cylinder, so that the pressure inside the cylinder is always at atmospheric pressure, although it is isolated from the atmosphere. In other words, the system is dynamically connected, by a movable boundary, to a constant-pressure reservoir.
- An isochoric (or isovolumetric) process is one in which the volume is held constant, meaning that the work done by the system will be zero. It follows that, for the simple system of two dimensions, any heat energy transferred to the system externally will be absorbed as internal energy. An isochoric process is also known as an isometric process. An example would be to place a closed tin can containing only air into a fire. To a first approximation, the can will not expand, and the only change will be that the gas gains internal energy, as evidenced by its increase in temperature and pressure. Mathematically, $\delta Q = dU$. We may say that the system is dynamically insulated, by a rigid boundary, from the environment

The temperature-entropy conjugate pair is concerned with the transfer of thermal energy as the result of heating.

- An isothermal process occurs at a constant temperature.

  An example would be to have a system immersed in a large constant-temperature bath. Any work energy performed by the system will be lost to the bath, but its temperature will remain constant. In other words, the system is thermally connected, by a thermally conductive boundary to a constant-temperature reservoir.

- An isentropic process occurs at a constant entropy. For a reversible process this is identical to an adiabatic process. If a system has an entropy which has not yet reached its maximum equilibrium value, a process of cooling may be required to maintain that value of entropy.
- An adiabatic process is a process in which there is no energy added or subtracted from the system by heating or cooling. For a reversible process, this is identical to an isentropic process. We may say that the system is thermally insulated from its environment and that its boundary is a thermal insulator. If a system has an entropy which has not yet reached its maximum equilibrium value, the entropy will increase even though the system is thermally insulated.

The above have all implicitly assumed that the boundaries are also impermeable to particles. We may assume boundaries that are both rigid and thermally insulating, but are permeable to one or more types of particle. Similar considerations then hold for the (chemical potential)-(particle number) conjugate pairs.

CONSERVATION OF ENERGY

In physics, the law of conservation of energy states that the total amount of energy in any isolated system remains constant but cannot be recreated, although it may change forms, e.g. friction turns kinetic energy into thermal energy. In thermodynamics, the first law of thermodynamics is a statement of the conservation of energy for thermodynamic systems, and is the more encompassing version of the conservation of energy. In short, the law of conservation of energy states that energy can not be created or destroyed, it can only be changed from one form to another or transferred from one body to another, but the total amount of energy remains constant (the same).

Ancient philosophers as far back as Thales of Miletus had inklings of the conservation of some underlying substance of which everything is made. However, there is no particular reason to identify this with what we know today as "mass-energy" (for example, Thales thought it was water) which can be described (in modern language) as conservatively converting potential energy to kinetic energy and back again. However, Galileo did not state the process in modern terms and again cannot be credited with the crucial insight.

It was Gottfried Wilhelm Leibniz during 1676–1689 who first attempted a mathematical formulation of the kind of energy which is connected with motion (kinetic energy). Leibniz noticed that in many mechanical systems (of several masses, mi each with velocity vi),

$$\sum_i m_i v_i^2$$

was conserved so long as the masses did not interact. He called this quantity the vis viva or living force of the system. The principle represents an accurate statement of the approximate conservation of kinetic energy in situations where there is no friction. Many physicists at that time were highly intelligent and held that the conservation of momentum, which holds even in systems with friction, as defined by the momentum:

$$\sum_i m_i v_i$$

was the conserved vis viva. It was later shown that, under the proper conditions, both quantities are conserved simultaneously such as in elastic collisions.

It was largely engineers such as John Smeaton, Peter Ewart, Karl Hotzmann, Gustave-Adolphe Hirn and Marc Seguin who objected that conservation of momentum alone was not adequate for practical calculation and who made use of Leibniz's principle.

The principle was also championed by some chemists such as William Hyde Wollaston. Academics such as John Playfair were quick to point out that kinetic energy is clearly not conserved. This is obvious to a modern analysis based on the second law of thermodynamics but in the 18th and 19th centuries, the fate of the lost energy was still unknown. Gradually it came to be suspected that the heat inevitably generated by motion under friction, was another form of vis viva.

In 1783, Antoine Lavoisier and Pierre-Simon Laplace reviewed the two competing theories of vis viva and caloric theory. Count Rumford's 1798 observations of heat generation during the boring of cannons added more weight to the view that mechanical motion could be converted into heat, and (as importantly) that the conversion was quantitative and could be predicted (allowing for a universal conversion constant between kinetic energy and heat). Vis viva now started to be known as energy, after the term was first used in that sense by Thomas Young in 1807.

The recalibration of vis viva to

$$\frac{1}{2}\sum_i m_i v_i^2$$

which can be understood as finding the exact value for the kinetic energy to work conversion constant, was largely the result of the work of Gaspard-Gustave Coriolis and Jean-Victor Poncelet over the period 1819–1839. The former called the quantity quantité de travail (quantity of work) and the latter, travail mécanique (mechanical work), and both championed its use in engineering calculation. It may appear, according to circumstances, as motion, chemical affinity, cohesion, electricity, light and magnetism; and from any one of these forms it can be transformed into any of the others."

A key stage in the development of the modern conservation principle was the demonstration of the mechanical equivalent of heat. The caloric theory maintained that heat could neither be created nor destroyed but conservation of energy entails the contrary principle that heat and mechanical work are interchangeable.

The mechanical equivalence principle was first stated in its modern form by the German surgeon Julius Robert von Mayer. Mayer reached his conclusion on a voyage to the Dutch

East Indies, where he found that his patients' blood was a deeper red because they were consuming less oxygen, and therefore less energy, to maintain their body temperature in the hotter climate. He had discovered that heat and mechanical work were both forms of energy, and later, after improving his knowledge of physics, he calculated a quantitative relationship between them.

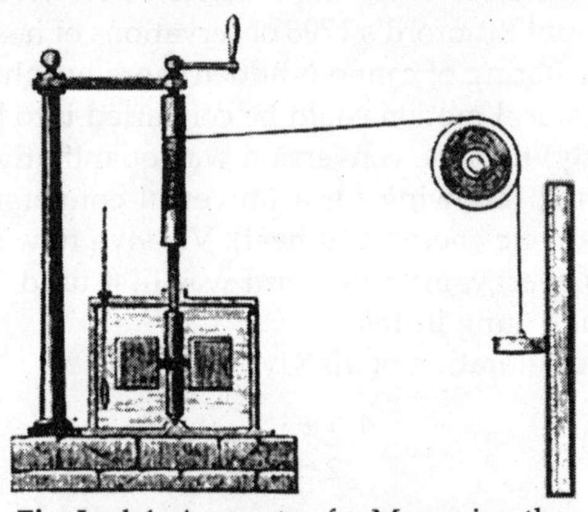

Fig. Joule's Apparatus for Measuring the Mechanical Equivalent of Heat.

Meanwhile, in 1843 James Prescott Joule independently discovered the mechanical equivalent in a series of experiments. In the most famous, now called the "Joule apparatus", a descending weight attached to a string caused a paddle immersed in water to rotate. He showed that the gravitational potential energy lost by the weight in descending was equal to the thermal energy (heat) gained by the water by friction with the paddle.

Over the period 1840–1843, similar work was carried out by engineer Ludwig A. Colding though it was little known outside his native Denmark.

Both Joule's and Mayer's work suffered from resistance and neglect but it was Joule's that, perhaps unjustly, eventually drew the wider recognition.In 1844, William Robert Grove postulated a relationship between mechanics, heat, light, electricity and magnetism by treating them all as manifestations of a single "force" (energy in modern terms).

THE FIRST LAW OF THERMODYNAMICS

Entropy is a function of a quantity of heat which shows the possibility of conversion of that heat into work.

For a thermodynamic system with a fixed number of particles, the first law of thermodynamics may be stated as:

$$\delta Q = dU + \delta W \text{, or equivalently, } dU = \delta Q - \delta W,$$

where δQ is the amount of energy added to the system by a heating process, δW is the amount of energy lost by the system due to work done by the system on its surroundings and dU is the increase in the internal energy of the system.

The δ's before the heat and work terms are used to indicate that they describe an increment of energy which is to be interpreted somewhat differently than the dU increment of internal energy.

Work and heat are processes which add or subtract energy, while the internal energy U is a particular form of energy associated with the system. Thus the term "heat energy" for δQ means "that amount of energy added as the result of heating" rather than referring to a particular form of energy.

Likewise, the term "work energy" for δW means "that amount of energy lost as the result of work". The most significant result of this distinction is the fact that one can clearly state the amount of internal energy possessed by a thermodynamic system, but one cannot tell how much energy has flowed into or out of the system as a result of its being heated or cooled, nor as the result of work being performed on or by the system. In simple terms, this means that energy cannot be created or destroyed, only converted from one form to another.

For a simple compressible system, the work performed by the system may be written

$$\delta Q = P\, dV,$$

where P is the pressure and dV is a small change in the volume of the system, each of which are system variables. The heat energy may be written

$$\delta Q = T\, dS,$$

where T is the temperature and dS is a small change in the entropy of the system. Temperature and entropy are also system variables.

MECHANICS

In mechanics, conservation of energy is usually stated as

$$E = T + V,$$

where T is kinetic and V potential energy.

Actually this is the particular case of the more general conservation law

$$\sum_{i=1}^{N} p_i \dot{q}_i - L = const \text{ and } p_i = \frac{\partial L}{\partial \dot{q}_i}$$

where L is the Lagrangian function. For this particular form to be valid, the following must be true:

- The system is scleronomous (neither kinetic nor potential energy are explicit functions of time)
- The kinetic energy is a quadratic form with regard to velocities.
- The potential energy doesn't depend on velocities.

Noether's Theorem

in many physical theories. It is understood as a consequence of Noether's theorem, which states every symmetry of a physical theory has an associated conserved quantity; if the theory's symmetry is time invariance then the conserved quantity is called "energy". In other words, if the theory is invariant under the continuous symmetry of time translation then its energy (which is canonical conjugate quantity to time) is conserved.

Conversely, theories which are not invariant under shifts in time (for example, systems with time dependent potential energy) do not exhibit conservation of energy — unless we consider them to exchange energy with another, external system so that the theory of the enlarged system becomes time invariant again. Since any time-varying theory can be embedded within a time-invariant meta-theory energy conservation can always be recovered by a suitable redefinition of what energy is. Thus conservation of energy is valid in all modern physical theories, such as special and general relativity and quantum theory (including QED).

Relativity

With the invention of special relativity by Albert Einstein, energy was proposed to be one component of an energy-momentum 4-vector. Each of the four components (one of energy and three of momentum) of this vector is separately conserved in any given inertial reference frame. Also conserved is the vector length (Minkowski norm), which is the rest mass.

The relativistic energy of a single massive particle contains a term related to its rest mass in addition to its kinetic energy of motion. In the limit of zero kinetic energy (or equivalently in the rest frame of the massive particle, or the center-of-momentum frame for objects or systems), the total energy of particle or object (including internal kinetic energy in systems) is related to its rest mass via the famous equation E = mc2.

Thus, the rule of conservation of energy in special relativity was shown to be a special case of a more general rule, alternatively called the conservation of mass and energy, the conservation of mass-energy, the conservation of energy-momentum, the conservation of invariant mass or now usually just referred to as conservation of energy.

In general relativity conservation of energy-momentum is expressed with the aids of a stress-energy-momentum pseudotensor.

Quantum Theory

In quantum mechanics, energy is defined as proportional to the time derivative of the wave function. Lack of commutation of the time derivative operator with the time operator itself mathematically results in an uncertainty principle for time and energy: the longer the period of time, the more precisely energy can be defined (energy and time become a conjugate Fourier pair). However, there is a deep contradiction between quantum theory's historical estimate of the vacuum energy density in the universe and the vacuum energy predicted by the cosmological constant.

The estimated energy density difference is of the order of 10^120 times. The consensus is developing that the quantum mechanical derived zero-point field energy density does not conserve the total energy of the universe, and does not comply with our understanding of the expansion of the universe. Intense effort is going on behind the scenes in physics to resolve this dilemma and to bring it into compliance with an expanding universe.

Mathematical Viewpoint

From a mathematical point of view, the energy conservation law is a consequence of the shift symmetry of time; energy conservation is implied by the empirical fact that the laws of physics do not change with time itself. Philosophically this can be stated as "nothing depends on time per se".

SECOND LAW OF THERMODYNAMICS

The second law of thermodynamics is an expression of the universal law of increasing entropy, stating that the entropy of an isolated system which is not in equilibrium will tend to increase over time, approaching a maximum value at equilibrium.

The second law traces its origin to French physicist Sadi Carnot's 1824 paper Reflections on the Motive Power of Fire, which presented the view that motive power (work) is due to the fall of caloric (heat) from a hot to cold body (working substance). In simple terms, the second law is an expression of the fact that over time, ignoring the effects of self-gravity, differences in temperature, pressure, and density tend to even out in a physical system that is isolated from the outside world. Entropy is a measure of how far along this evening-out process has progressed.

There are many versions of the second law, but they all have the same effect, which is to explain the phenomenon of irreversibility in nature.

Versions of The Law

There are many ways of stating the second law of thermodynamics, but all are equivalent in the sense that each form of the second law logically implies every other form. Thus, the theorems of thermodynamics can be proved using any form of the second law and third law

The formulation of the second law that refers to entropy directly is as follows:

In a system, a process can occur only if it increases the total entropy of the universe.

Thus, while a system can undergo some physical process that decreases its own entropy, the entropy of the universe (which includes the system and its surroundings) must increase overall. (An exception to this rule is a reversible or "isentropic" process, such as frictionless adiabatic compression.) Processes that decrease total entropy of the universe are impossible. If a system is at equilibrium, by definition no spontaneous processes occur, and therefore the system is at maximum entropy.

Also, due to Rudolf Clausius, is the simplest formulation of the second law, the heat formulation or Clausius statement:

Heat cannot spontaneously flow from a material at lower temperature to a material at higher temperature.

Informally, "Heat doesn't flow from cold to hot (without work input)", which is obviously true from everyday experience. For example in a refrigerator, heat flows from cold to hot, but only when aided by an external agent, i.e. the compressor.

Note that from the mathematical definition of entropy, a process in which heat flows from cold to hot has decreasing entropy. This is allowable in a non-isolated system, however only if entropy is created elsewhere, such that the total entropy is constant or increasing, as required by the second law. For example, the electrical energy going into a refrigerator is converted to heat and goes out the back, representing a net increase in entropy.

A third formulation of the second law, by Lord Kelvin, is the heat engine formulation, or Kelvin statement:

It is impossible to convert heat completely into work.

That is, it is impossible to extract energy by heat from a high-temperature energy source and then convert all of the energy into work. At least some of the energy must be passed on to heat a low-temperature energy sink. Thus, a heat engine with 100% efficiency is thermodynamically impossible.

Microscopic Systems

Thermodynamics is a theory of macroscopic systems at equilibrium and therefore the second law applies only to macroscopic systems with well-defined temperatures. On scales of a few atoms, the second law does not apply; for example, in a system of two molecules, it is possible for the slower-moving ("cold") molecule to transfer energy to the faster-moving ("hot") molecule. Such tiny systems are outside the domain of classical thermodynamics, but they can be investigated in quantum thermodynamics by using statistical mechanics. For any isolated system with a mass of more than a few picograms, the second law is true to within a few parts in a million.

Energy Dispersal

The second law of thermodynamics is an axiom of thermodynamics concerning heat, entropy, and the direction in which thermodynamic processes can occur. For example, the second law implies that heat does not spontaneously flow from a cold material to a hot material, but it allows heat to flow from a hot material to a cold material. Roughly speaking, the second law says that in an isolated system, concentrated energy disperses over time, and consequently less concentrated energy is available to do useful work. Energy dispersal also means that differences in temperature, pressure, and density even out.

Again roughly speaking, thermodynamic entropy is a measure of energy dispersal, and so the second law is closely connected with the concept of entropy. The second law says that temperature differences between systems in contact with each other tend to even out and that work can be obtained from these non-equilibrium differences, but that loss of heat occurs, in the form of entropy, when work is done. Pressure differences, density differences, and particularly temperature differences, all tend to equalize if given the opportunity. This means that an isolated system will eventually come to have a uniform temperature. A heat engine is a mechanical device that provides useful work from the difference in temperature of two bodies:

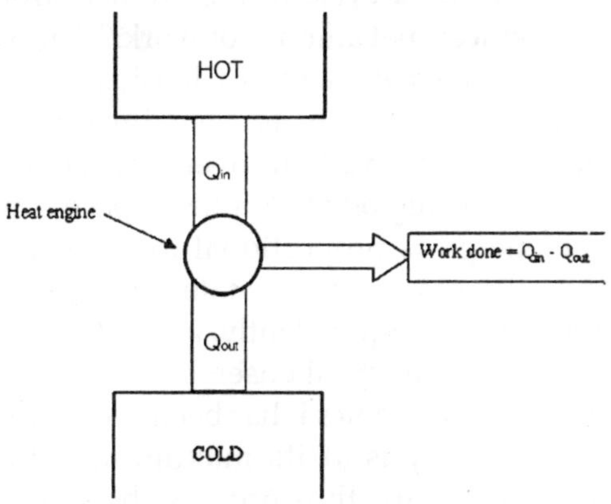

Fig. Heat Engine Diagram

During the 19th century, the second law was synthesized, essentially, by studying the dynamics of the Carnot heat engine in coordination with James Joule's Mechanical equivalent of heat experiments. Since any thermodynamic engine requires such a temperature difference, it follows that no useful work can be derived from an isolated system in equilibrium; there must always be an external energy source and a cold sink. By definition, perpetual motion machines of the second kind would have to violate the second law to function.

The first theory on the conversion of heat into mechanical work is due to Nicolas Léonard Sadi Carnot in 1824. He was the first to realize correctly that the efficiency of this conversion depends on the difference of temperature between an engine and its environment.

Recognizing the significance of James Prescott Joule's work on the conservation of energy, Rudolf Clausius was the first to formulate the second law in 1850, in this form: heat does not spontaneously flow from cold to hot bodies. While common knowledge now, this was contrary to the caloric theory of heat popular at the time, which considered heat as a liquid. From there he was able to infer the law of Sadi Carnot and the definition of entropy .

Established in the 19th century, the Kelvin-Planck statement of the Second Law says, "It is impossible for any device that operates on a cycle to receive heat from a single reservoir and produce a net amount of work." This was shown to be equivalent to the statement of Clausius.

The Ergodic hypothesis is also important for the Boltzmann approach. It says that, over long periods of time, the time spent in some region of the phase space of microstates with the same energy is proportional to the volume of this region, i.e. that all accessible microstates are equally probable over long period of time. Equivalently, it says that time average and average over the statistical ensemble are the same.

Using quantum mechanics it has been shown that the local von Neumann entropy is at its maximum value with an extremely high probability, thus proving the second law. The result is valid for a large class of isolated quantum systems (e.g. a gas in a container). While the full system is pure and has therefore no entropy, the entanglement between gas and container gives rise to an increase of the local entropy of the gas. This result is one of the most important achievements of quantum thermodynamics.

Informal Descriptions

The second law can be stated in various succinct ways, including:

- It is impossible to produce work in the surroundings using a cyclic process connected to a single heat reservoir (Kelvin, 1851).
- It is impossible to carry out a cyclic process using an engine connected to two heat reservoirs that will have as its only effect the transfer of a quantity of heat from the low-temperature reservoir to the high-temperature reservoir (Clausius, 1854).

- If thermodynamic work is to be done at a finite rate, free energy must be expended.

MATHEMATICAL DESCRIPTIONS

In 1856, the German physicist Rudolf Clausius stated what he called the "second fundamental theorem in the mechanical theory of heat" in the following form:

$$\int \frac{\delta Q}{T} = -N$$

where N is the "equivalence-value" of all uncompensated transformations involved in a cyclical process. Later, in 1865, Clausius would come to define "equivalence-value" as entropy. On the heels of this definition, that same year, the most famous version of the second law was read in a presentation at the Philosophical Society of Zurich on April 24th, in which, in the end of his presentation, Clausius concludes:

The entropy of the universe tends to a maximum.

This statement is the best-known phrasing of the second law. Moreover, owing to the general broadness of the terminology used here, e.g. universe, as well as lack of specific conditions, e.g. open, closed, or isolated, to which this statement applies, many people take this simple statement to mean that the second law of thermodynamics applies virtually to every subject imaginable. This, of course, is not true; this statement is only a simplified version of a more complex description.

In terms of time variation, the mathematical statement of the second law for a closed system undergoing an arbitrary transformation is:

$$\frac{dS}{dt} > 0$$

where

S is the entropy and

t is time.

It should be noted that statistical mechanics gives an explanation for the second law by postulating that a material is composed of atoms and molecules which are in constant motion.

A particular set of positions and velocities for each particle in the system is called a microstate of the system and because of the constant motion, the system is constantly changing its microstate. Statistical mechanics postulates that, in equilibrium, each microstate that the system might be in is equally likely to occur, and when this assumption is made, it leads directly to the conclusion that the second law must hold in a statistical sense.

That is, the second law will hold on average, with a statistical variation on the order of 1/✓N where N is the number of particles in the system. For everyday (macroscopic) situations, the probability that the second law will be violated is practically nil. However, for systems with a small number of particles, thermodynamic parameters, including the entropy, may show significant statistical deviations from that predicted by the second law. Classical thermodynamic theory does not deal with these statistical variations.

AVAILABLE USEFUL WORK

An important and revealing idealized special case is to consider applying the Second Law to the scenario of an isolated system (called the total system or universe), made up of two parts: a sub-system of interest, and the sub-system's surroundings.

These surroundings are imagined to be so large that they can be considered as an unlimited heat reservoir at temperature TR and pressure PR — so that no matter how much heat is transferred to (or from) the sub-system, the temperature of the surroundings will remain TR; and no matter how much the volume of the sub-system expands (or contracts), the pressure of the surroundings will remain PR.

Whatever changes dS and dSR occur in the entropies of the sub-system and the surroundings individually, according to the Second Law the entropy Stot of the isolated total system must not decrease:

$$dS_{\text{tot}} = dS + dS_R \geq 0.$$

According to the First Law of Thermodynamics, the change dU in the internal energy of the sub-system is the sum of the heat δq added to the sub-system, less any work δw done by the sub-system, plus any net chemical energy entering the sub-system d ΣμiRNi, so that:

$$dS = \delta q - \delta w + d\left(\sum \mu_{iR} N_i\right)$$

where μiR are the chemical potentials of chemical species in the external surroundings.

Now the heat leaving the reservoir and entering the sub-system is

$$\delta q = T_R(-dS_R) \leq T_R dS$$

where we have first used the definition of entropy in classical thermodynamics (alternatively, the definition of temperature in statistical thermodynamics); and then the Second Law inequality from above.

It therefore follows that any net work δw done by the sub-system must obey

$$\delta w \leq -dU + T_R dS + \sum \mu_{iR} dN_i.$$

It is useful to separate the work δw done by the subsystem into the useful work δwu that can be done by the sub-system, over and beyond the work pR dV done merely by the sub-system expanding against the surrounding external pressure, giving the following relation for the useful work that can be done:

$$\delta w_u \leq -d(U - T_R S + p_R V - \sum \mu_{iR} N_i).$$

It is convenient to define the right-hand-side as the exact derivative of a thermodynamic potential, called the availability or exergy X of the subsystem,

$$X = U - T_R S + p_R V - \sum \mu_{iR} N_i$$

The Second Law therefore implies that for any process which can be considered as divided simply into a subsystem, and an unlimited temperature and pressure reservoir with which it is in contact,

$$dX + \delta w_u \leq 0$$

i.e. the change in the subsystem's exergy plus the useful work done by the subsystem (or, the change in the subsystem's exergy less any work, additional to that done by the pressure reservoir, done on the system) must be less than or equal to zero.

Special Cases: Gibbs and Helmholtz Free Energies

When no useful work is being extracted from the sub-system, it follows that

$$dX \leq 0$$

with the exergy X reaching a minimum at equilibrium, when dX=0.

If no chemical species can enter or leave the sub-system, then the term ΣμiR Ni can be ignored. If furthermore the temperature of the sub-system is such that T is always equal to TR, then this gives:

$$X = U - TS + p_R V + \text{const}.$$

If the volume V is constrained to be constant, then

$$X = U - TS + \text{const}' = A + \text{const}'$$

where A is the thermodynamic potential called Helmholtz free energy, A=U–TS. Under constant volume conditions therefore, dA d" 0 if a process is to go forward; and dA=0 is the condition for equilibrium.

Alternatively, if the sub-system pressure p is constrained to be equal to the external reservoir pressure pR, then

$$X = U - TS + pV + \text{const}. = G + \text{const}.$$

where G is the Gibbs free energy, G=U–TS+PV. Therefore under constant pressure conditions, if dG d ≤ 0, then the process can occur spontaneously, because the change in system energy exceeds the energy lost to entropy. dG=0 is the condition for equilibrium. This is also commonly written in terms of enthalpy, where H=U+PV.

Application

In sum, if a proper infinite-reservoir-like reference state is chosen as the system surroundings in the real world, then the Second Law predicts a decrease in X for an irreversible process and no change for a reversible process.

$$dS_{tot} \geq 0 \text{ is equivalent to } dX + \delta W_u \leq 0$$

This expression together with the associated reference state permits a design engineer working at the macroscopic scale (above the thermodynamic limit) to utilize the Second Law without directly measuring or considering entropy change in a total isolated system.

Those changes have already been considered by the assumption that the system under consideration can reach equilibrium with the reference state without altering the reference state. An efficiency for a process or collection of processes that compares it to the reversible ideal may also be found.

This approach to the Second Law is widely utilized in engineering practice, environmental accounting, systems ecology, and other disciplines.

CRITICISMS

Owing to the somewhat ambiguous nature of the formulation of the second law, i.e. the postulate that the quantity heat divided by temperature increases in spontaneous natural processes, it has occasionally been subject to criticism as well as attempts to dispute or disprove it.

Clausius himself even noted the abstract nature of the second law. In his 1862 memoir, for example, after mathematically stating the second law by saying that integral of the differential of a quantity of heat divided by temperature must be greater than or equal to zero for every cyclical process which is in any way possible:

$$\oint \frac{\delta Q}{T} \geq 0 .$$

Clausius then stated: Although the necessity of this theorem admits of strict mathematical proof if we start from the fundamental proposition above quoted it thereby nevertheless retains an abstract form, in which it is with difficulty embraced by the mind, and we feel compelled to seek for the precise physical cause, of which this theorem is a consequence.

Recall that heat and temperature are statistical, macroscopic quantities that become somewhat ambiguous when dealing with a small number of atoms.

Perpetual Motion of the Second Kind

Before 1850, heat was regarded as an indestructible particle of matter. This was called the "material hypothesis", as based principally on the views of Isaac Newton. It was on these views, partially, that in 1824 Sadi Carnot formulated the initial version of the second law.

It soon was realized, however, that if the heat particle was conserved, and as such not changed in the cycle of an engine, that it would be possible to send the heat particle cyclically through the working fluid of the engine and use it to push the piston and then return the particle, unchanged, to its original state. In this manner perpetual motion could be created and used as an unlimited energy source. Thus, historically, people have always been attempting to create a perpetual motion machine so to disprove the second law.

Maxwell's Demon

In 1871, James Clerk Maxwell proposed a thought experiment, now called Maxwell's demon, which challenged the second law. This experiment reveals the importance of observability in discussing the second law. In other words, it requires a certain amount of energy to collect the information necessary for the demon to "know" the whereabouts of all the particles in the system. This energy requirement thus negates the challenge to the second law. Moreover, to reconcile this apparent paradox from another perspective, one may resort to a use of information entropy, although this is considered questionable by some.

Time's Arrow

The second law is a law about macroscopic irreversibility, and so may appear to violate the principle of T-symmetry. Boltzmann first investigated the link with microscopic reversibility. In his H-theorem he gave an explanation, by means of statistical mechanics, for dilute gases in the zero density limit where the ideal gas equation of state holds.

He derived the second law of thermodynamics not from mechanics alone, but also from the probability arguments. His idea was to write an equation of motion for the probability that a single particle has a particular position and momentum at a particular time. One of the terms in this equation accounts for how the single particle distribution changes through collisions of pairs of particles.

This rate depends of the probability of pairs of particles. Boltzmann introduced the assumption of molecular chaos to reduce this pair probability to a product of single particle probabilities. From the resulting Boltzmann equation he derived his famous H-theorem, which implies that on average the entropy of an ideal gas can only increase.

The assumption of molecular chaos in fact violates time reversal symmetry. It assumes that particle momenta are uncorrelated before collisions. If you replace this assumption with "anti-molecular chaos," namely that particle momenta are uncorrelated after collision, then you can derive an anti-Boltzmann equation and an anti-H-Theorem which implies entropy decreases on average. Thus we see that in reality Boltzmann did not succeed in solving Loschmidt's paradox. The molecular chaos assumption is the key element that introduces the arrow of time.

Applications to Living Systems

The second law of thermodynamics has been proven mathematically for thermodynamic systems, where entropy is defined in terms of heat divided by the absolute temperature. The second law is often applied to other situations, such as the complexity of life, or orderliness. In sciences such as biology and biochemistry the application of thermodynamics is well-established, e.g. biological thermodynamics.

The general viewpoint on this subject is summarized well by biological thermodynamicist Donald Haynie; as he states: "Any theory claiming to describe how organisms originate and continue to exist by natural causes must be compatible with the first and second laws of thermodynamics."

This is very different, however, from the claim made by many creationists that evolution violates the second law of thermodynamics. In fact, evidence indicates that biological systems and obviously the evolution of those systems conform to the second law, since although biological systems may become more ordered, the net increase in entropy for the entire universe is still positive as a result of evolution.

Complex Systems

It is occasionally claimed that the second law is incompatible with autonomous self-organisation, or even the coming into existence of complex systems. This is a common creationist argument against evolution. The entry self-organisation explains how this claim is a misconception. In fact, as hot systems cool down in accordance with the second law, it is not unusual for them to undergo spontaneous symmetry breaking, i.e. for structure to spontaneously appear as the temperature drops below a critical threshold. Complex structures, such as Bénard cells, also spontaneously appear where there is a steady flow of energy from a high temperature input source to a low temperature external sink.

Furthermore, a system that energy flows into and out of may decrease its local entropy provided the increase of the entropy to its surrounding that this process causes is greater than or equal to the local decrease in entropy. A good example of this is crystallization as a liquid cools, crystals begin to form inside it. While these crystals are more ordered than the liquid they originated from, in order for them to form they must release a great deal of heat, known as the latent heat of fusion. This heat flows out of the system and increases the entropy of its surroundings to a greater extent than the decrease of energy that the liquid undergoes in the formation of crystals.

An interesting situation to consider is that of a supercooled liquid perfectly isolated thermodynamically, into which a grain of dust is dropped. Here even though the system cannot export energy to its surroundings, it will still crystallize. Now however the release of latent heat will contribute to raising its own temperature. If this release of heat causes the temperature to reach the melting point before it has fully crystallized, then it shall remain a mixture of liquid and solid; if not, then it will be a solid at a significantly higher temperature than it previously was as a liquid. In both cases entropy from its disordered structure is converted into entropy of disordered motion.

THE THIRD LAW OF THERMODYNAMICS

The third law of thermodynamics is usually stated as a definition: the entropy of a perfect crystal of an element at the absolute zero of temperature is zero.

At the absolute zero of temperature, there is zero thermal energy or heat. Since heat is a measure of average molecular motion, zero thermal energy means that the average atom does not move at all. Since no atom can have less than zero motion, the motion of every individual atom must be zero when the average molecular motion is zero. When none of the atoms which make up a perfectly ordered crystal move at all, there can be no disorder or different states possible for the crystal.

Taking the entropy of a perfect crystal of an element to be zero at the absolute zero of temperature establishes a method by which entropies of elements at any higher temperature can be determined. Since by definition dS = qrev/T, the mathematical integral of dS from zero to any higher temperature T is the integral, over that temperature range, of qrev/T. In other words. the difference S - S0 is the integral from zero to a temperature T of (Cp/T)dT. The molar heat capacity at any temperature is a measurable quantity, and so this difference can be determined experimentally.

The third law of thermodynamics is simply the statement that S0 is zero by definition for a pure element, and so if the heat capacity is measured under conditions of reversible heat flow, as it can be, and as a function of temperature at low temperatures, as it can be, then the entropy Sof a pure element at any temperature T is given by:

S = the integral from zero to T of (Cp/T)dT

The values we usually write as S0, the standard entropy of a substance, are actually the integral from zero to the standard temperature, 298.15 K, of (Cp/T)dT in order that the values of standard entropies of substances can be used with standard values for other thermodynamic functions such as enthalpy and free energy.

The value of the entropy of an element at any temperature, including 298.15 K which is the temperature of the standard entropy S0, can be obtained from careful measurements of the heat capacity of the element from the desired temperature down to absolute zero. Experimentally, chemists have been unable to reach the absolute zero of temperature, but measurements can be and have been made down to within 0.1 K of it and the heat capacity below measurable range can be accurately estimated. The entropy of the element is obtained from integration of the heat capacity measurements. If the crystal changes form, or melts, within the temperature range desired, then the entropy of that change must also be measured and included.

The entropies of compounds can be determined from thermodynamic measurements made on the reactions which form them from the elements or dissociate them to their constituent elements once the entropies of the elements themselves are established. The values of standard entropies of elements and compounds at 25°C of the thermodynamic properties of pure substances are all based upon the third law of thermodynamics.

ADIABATIC PROCESS

In thermodynamics, an adiabatic process or an isocaloric process is a thermodynamic process in which no heat is transferred to or from the working fluid. The term "adiabatic" literally means impassable (from Greek -äép-âáÖíåéí, not-through-to pass), corresponding here to an absence of heat transfer. Conversely, a process that involves heat transfer (addition or loss of heat to the surroundings) is generally called diabatic. Linguistically, diabatic is the opposite of adiabatic by the absence of the initial a.

For example, an adiabatic boundary is a boundary that is impermeable to heat transfer and the system is said to be adiabatically (or thermally) insulated; an insulated wall approximates an adiabatic boundary. Another example is the adiabatic flame temperature, which is the temperature that would be achieved by a flame in the absence of heat loss to the surroundings. An adiabatic process that is reversible is also called an isentropic process. Additionally, an adiabatic process that is irreversible and extracts no work is in an isenthalpic process, such as viscous drag, progressing towards a nonnegative change in entropy.

One opposite extreme—allowing heat transfer with the surroundings, causing the temperature to remain constant—is known as an isothermal process. Since temperature is thermodynamically conjugate to entropy, the isothermal process is conjugate to the adiabatic process for reversible transformations.

A transformation of a thermodynamic system can be considered adiabatic when it is quick enough that no significant heat is transferred between the system and the outside. At the opposite extreme, a transformation of a thermodynamic system can be considered isothermal if it is slow enough so that the system's temperature remains constant by heat exchange with the outside.

ADIABATIC HEATING AND COOLING

Adiabatic changes in temperature occur due to changes in pressure of a gas while not adding or subtracting any heat. Adiabatic heating occurs when the pressure of a gas is increased from work done on it by its surroundings, ie a piston. Diesel engines rely on adiabatic heating during their compression stroke to elevate the temperature sufficiently to ignite the fuel. Similarly jet engines rely upon adiabatic heating to create the correct compression of the air to enable fuel to be injected and ignition to then occur.

Adiabatic heating also occurs in the Earth's atmosphere when an air mass descends, for example, in a katabatic wind or Foehn wind flowing downhill.

Adiabatic cooling occurs when the pressure of a substance is decreased as it does work on its surroundings. Adiabatic cooling does not have to involve a fluid. One technique used to reach very low temperatures (thousandths and even millionths of a degree above absolute zero) is adiabatic demagnetisation, where the change in magnetic field on a magnetic material is used to provide adiabatic cooling.

Adiabatic cooling also occurs in the Earth's atmosphere with orographic lifting and lee waves, and this can form pileus or lenticular clouds if the air is cooled below the dew point.

Rising magma also undergoes adiabatic cooling before eruption. Such temperature changes can be quantified using the ideal gas law, or the hydrostatic equation for atmospheric processes. It should be noted that no process is truly adiabatic. Many processes are close to adiabatic and can be easily approximated by using an adiabatic assumption, but there is always some heat loss. There is no such thing as a perfect insulator.

IDEAL GAS (REVERSIBLE CASE ONLY)

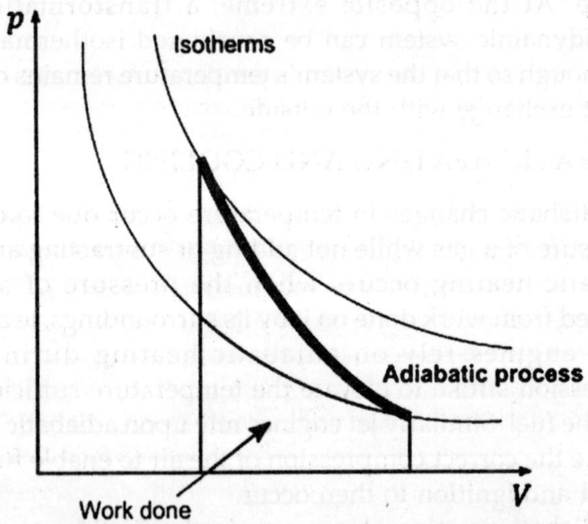

For a simple substance, during an adiabatic process in which the volume increases, the internal energy of the working substance must decrease

The mathematical equation for an ideal fluid undergoing a reversible (i.e., no entropy generation) adiabatic process is

$$PV^{\gamma} = \text{constant}$$

where P is pressure, V is volume, and

$$\gamma = \frac{C_p}{C_V} = \frac{\alpha + 1}{\alpha},$$

CP being the specific heat for constant pressure and CV being the specific heat for constant volume. α comes from the number of degrees of freedom divided by 2 (3/2 for monatomic gas, 5/2 for diatomic gas). For a monatomic ideal gas, $\gamma = 5 / 3$, and for a diatomic gas (such as nitrogen and oxygen, the main components of air) $\gamma = 7 / 5$. Note that the above formula is only applicable to classical ideal gases and not Bose-Einstein or Fermi gases.

For reversible adiabatic processes, it is also true that

$$P^{\gamma-1} T^{-\gamma} = \text{constant}$$

$$VT^{\alpha} = \text{constant}$$

where T is an absolute temperature.

This can also be written as

$$TV^{\gamma-1} = \text{constant}$$

Derivation of Continuous Formula

The definition of an adiabatic process is that heat transfer to the system is zero, $\delta Q = 0$. Then, according to the first law of thermodynamics,

$$dU + \delta W = \delta W = 0,$$

where dU is the change in the internal energy of the system and δW is work done by the system.

Any work (δW) done must be done at the expense of internal energy U, since no heat δQ is being supplied from the surroundings. Pressure-volume work δW done by the system is defined as

$$\delta W = PdV.$$

However, P does not remain constant during an adiabatic process but instead changes along with V.It is desired to know how the values of dP and dV relate to each other as the adiabatic process proceeds. For an ideal gas the internal energy is given by

$$U = \alpha nRT,$$

where R is the universal gas constant and n is the number of moles in the system (a constant).

Differentiating Equation and use of the ideal gas law, $PV = nRT$, yields

$$dU = \alpha nRT = \alpha d(PV) = a(P dV + V\, dP).$$

Equation is often expressed as $dU = nC_V\, dT$ because CV = αR.

Now substitute equations and into equation to obtain

$$-P\, dV = \alpha P\, dV + \alpha V\, dP,$$

simplify:

$$-(\alpha + 1)P\, dV = \alpha P\, dV,$$

and divide both sides by PV:

$$-(\alpha + 1)\frac{dV}{V} = \alpha \frac{dP}{P}.$$

After integrating the left and right sides from V0 to V and from P0 to P and changing the sides respectively,

$$\ln\left(\frac{P}{P_0}\right) = -\frac{\alpha + 1}{\alpha}\ln\left(\frac{V}{V_0}\right).$$

Exponentiate both sides,

$$\left(\frac{P}{P_0}\right) = \left(\frac{V}{V_0}\right)^{-\frac{\alpha+1}{\alpha}},$$

and eliminate the negative sign to obtain

$$\left(\frac{P}{P_0}\right) = \left(\frac{V}{V_0}\right)^{-\frac{\alpha+1}{\alpha}}.$$

Therefore,

$$\left(\frac{P}{P_0}\right) = \left(\frac{V}{V_0}\right)^{\frac{\alpha+1}{\alpha}} = 1$$

and

$$PV^{\frac{\alpha+1}{\alpha}} = P_0 V_0^{\frac{\alpha+1}{\alpha}} = PV^{\gamma} = \text{constant}.$$

Derivation of Discrete Formula

The change in internal energy of a system, measured from state 1 to state 2, is equal to

$$\delta U = \alpha R n_2 T_2 - a R n_1 T_1 = \alpha R(n_2 T_2 - n_1 T_1)$$

At the same time, the work done by the pressure-volume changes as a result from this process, is equal to

$$\delta W = P_2 V_2 - P_1 V_1.$$

Since we require the process to be adiabatic, the following equation needs to be true

$$\delta U + \delta W = 0.$$

$$\alpha R(n_2 T_2 - n_1 T_1) + (P_2 V_2 - P_1 V_1) = 0$$

or

$$\frac{(P_2 V_2 - P_1 V_1)}{-(n_2 T_2 - n_1 T_1)} = \alpha R.$$

If it's further assumed that there are no changes in molar quantity (as often in practical cases), the formula is simplified to this one:

$$\frac{(P_2 V_2 - P_1 V_1)}{-(T_2 - T_1)} = \alpha n R.$$

GRAPHING ADIABATS

An adiabat is a curve of constant entropy on the P–V diagram. Properties of adiabats on a P–V diagram are:

- Every adiabat asymptotically approaches both the V axis and the P axis (just like isotherms).
- Each adiabat intersects each isotherm exactly once.
- An adiabat looks similar to an isotherm, except that during an expansion, an adiabat loses more pressure than an isotherm, so it has a steeper inclination (more vertical).

- If isotherms are concave towards the "north-east" direction (45°), then adiabats are concave towards the "east north-east" (31°).
- If adiabats and isotherms are graphed severally at regular changes of entropy and temperature, respectively (like altitude on a contour map), then as the eye moves towards the axes (towards the south-west), it sees the density of isotherms stay constant, but it sees the density of adiabats grow. The exception is very near absolute zero, where the density of adiabats drops sharply and they become rare.

The following diagram is a P–V diagram with a superposition of adiabats and isotherms:

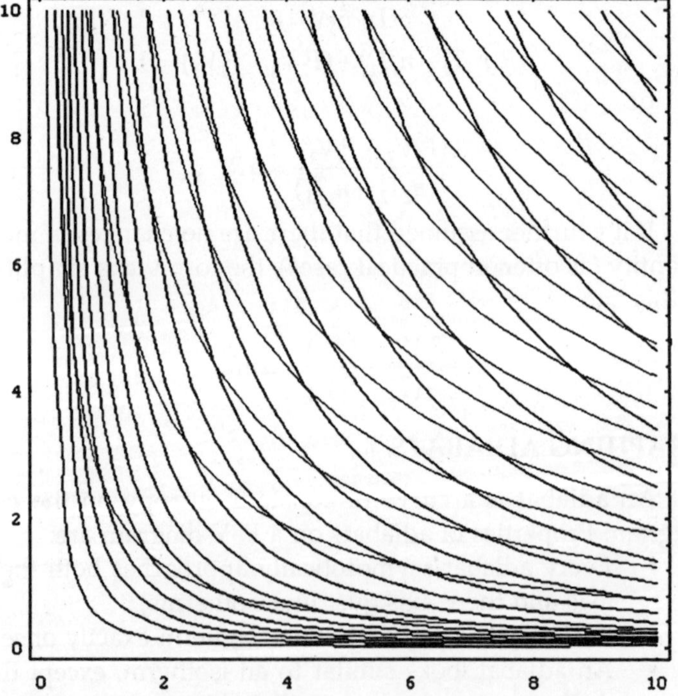

The isotherms are the red curves and the adiabats are the black curves. The adiabats are isentropic. Volume is the horizontal axis and pressure is the vertical axis.

Chapter 7

Quantum Mechanics

The statistical thermodynamics of ideal gases using a rather ad hoc combination of classical and quantum mechanics. In fact, we employed classical mechanics to deal with the translational degrees of freedom of the constituent particles, and quantum mechanics to deal with the non-translational degrees of freedom.

Let us now discuss ideal gases from a purely quantum mechanical standpoint. It turns out that this approach is necessary to deal with either low temperature or high density gases. Furthermore, it also allows us to investigate completely non-classical "gases," such as photons or the conduction electrons in a metal.

Symmetry Requirements in Quantum Mechanics

Consider a gas consisting of N identical, non-interacting, structureless particles enclosed within a container of volume V. Let Q_i denote collectively all the coordinates of the ith particle: i.e., the three Cartesian coordinates which determine its spatial position, as well as the spin coordinate which determines its internal state.

Let s_i be an index labeling the possible quantum states of the th particle: i.e., each possible value of s_i corresponds to a specification of the three momentum components of the particle, as well as the direction of its spin orientation. According to quantum mechanics, the overall state of the system when the th particle is in state s_i, etc., is completely determined by the complex wave-function

$$\Psi_{s_1,\dots,s_N}(Q_1, Q_2, \dots, Q_N).$$

In particular, the probability of an observation of the system finding the ith particle with coordinates in the range Q_i to $Q_i + dQ_i$, etc., is simply

$$|\Psi_{s_1,\ldots,s_N}(Q_1, Q_2, \ldots, Q_N)|^2 \, dQ_1 dQ_2 \ldots dQ_N.$$

One of the fundamental postulates of quantum mechanics is the essential indistinguishability of particles of the same species. What this means, in practice, is that we cannot label particles of the same species: i.e., a proton is just a proton—we cannot meaningfully talk of proton number 1 and proton number 2, etc. Note that no such constraint arises in classical mechanics. Thus, in classical mechanics particles of the same species are regarded as being distinguishable, and can, therefore, be labelled. Of course, the quantum mechanical approach is the correct one.

Suppose that we interchange the ith and jth particles: i.e.,

$$Q_i \leftrightarrow Q_j$$

$$s_i \leftrightarrow s_j.$$

If the particles are truly indistinguishable then nothing has changed: i.e., we have a particle in quantum state s and a particle in quantum state s_j both before and after the particles are swapped. Thus, the probability of observing the system in a given state also cannot have changed: i.e.,

$$\left|\Psi(Q_i \cdots Q_j \cdots)\right|^2 = \left|\Psi(\cdots Q_j \cdots Q_i \cdots)\right|^2.$$

Here, we have omitted the subscripts $s_1, \ldots, s_N$ for the sake of clarity. Note that we cannot conclude that the wave-function Ψ is unaffected when the particles are swapped, because Ψ cannot be observed experimentally. Only the probability density $|\Psi|^2$ is observable. Equation implies that

$$\Psi(\cdots Q_i \cdots Q_j \cdots) = A\Psi(\cdots Q_i \cdots Q_j \cdots),$$

where A is a complex constant of modulus unity: i.e., $|A|^2 = 1$.

Suppose that we interchange the ith and jth particles a second time. Swapping the ith and jth particles twice leaves the system completely unchanged: i.e., it is equivalent to doing nothing to the system. Thus, the wave-functions before and after this process must be identical. It follows from Eq. that

$$A^2 = 1$$

Of course, the only solutions to the above equation are $A = \pm 1$.

We conclude, from the above discussion, that the wave-function Ψ is either completely symmetric under the interchange of particles, or it is completely anti-symmetric. In other words, either

$$\Psi(\cdots Q_i \cdots Q_j \cdots) = +\Psi(\cdots Q_j \cdots Q_i \cdots),$$

or

$$\Psi(\cdots Q_i \cdots Q_j \cdots) = -\Psi(\cdots Q_j \cdots Q_i \cdots).$$

In 1940 the Nobel prize winning physicist Wolfgang Pauli demonstrated, via arguments involving relativistic invariance, that the wave-function associated with a collection of identical integer-spin (i.e., spin 0, 1, 2, etc.) particles satisfies Eq. whereas the wave-function associated with a collection of identical half-integer-spin (i.e., spin 1/2,3/2,5/2, etc.) particles satisfies Eq. The former type of particles are known as bosons [after the Indian physicist S.N. Bose, who first put forward Eq. on empirical grounds].

The latter type of particles are called fermions (after the Italian physicists Enrico Fermi, who first studied the properties of fermion gases). Common examples of bosons are photons and He^4 atoms. Common examples of fermions are protons, neutrons, and electrons.

Consider a gas made up of identical bosons. Equation implies that the interchange of any two particles does not lead to a new state of the system. Bosons must, therefore, be considered as genuinely indistinguishable when enumerating the different possible states of the gas. Note that Eq. imposes no restriction on how many particles can occupy a given single-particle quantum state *s*.

Consider a gas made up of identical fermions. Equation implies that the interchange of any two particles does not lead to a new physical state of the system (since $|\Psi|^2$ is invariant). Hence, fermions must also be considered genuinely indistinguishable when enumerating the different possible states of the gas.

Consider the special case where particles i and j lie in the same quantum state. In this case, the act of swapping the two particles is equivalent to leaving the system unchanged, so

$$\Psi(\cdots Q_i \cdots Q_j \cdots) = \Psi(\cdots Q_j \cdots Q_i \cdots).$$

However, Eq. is also applicable, since the two particles are fermions. The only way in which Eqs. and can be reconciled is if

$$\Psi = 0$$

wherever particles i and j lie in the same quantum state. This is another way of saying that it is impossible for any two particles in a gas of fermions to lie in the same single-particle quantum state. This proposition is known as the Pauli exclusion principle, since it was first proposed by W. Pauli in 1924 on empirical grounds.

Consider, for the sake of comparison, a gas made up of identical classical particles. In this case, the particles must be considered distinguishable when enumerating the different possible states of the gas. Furthermore, there are no constraints on how many particles can occupy a given quantum state.

According to the above discussion, there are three different sets of rules which can be used to enumerate the states of a gas made up of identical particles. For a boson gas, the particles must be treated as being indistinguishable, and there is no limit to how many particles can occupy a given quantum state. This set of rules is called Bose-Einstein statistics, after S.N. Bose and A. Einstein, who first developed them. For a fermion gas, the particles must be treated as being indistinguishable, and there can never be more than one particle in any given quantum state.

This set of rules is called Fermi-Dirac statistics, after E. Fermi and P.A.M. Dirac, who first developed them. Finally, for a classical gas, the particles must be treated as being distinguishable, and there is no limit to how many particles can occupy a given quantum state.

This set of rules is called Maxwell-Boltzmann statistics, after J.C. Maxwell and L. Boltzmann, who first developed them.

An Illustrative Example

Consider a very simple gas made up of two identical particles. Suppose that each particle can be in one of three possible quantum states, s = 1, 2, 3. Let us enumerate the possible states of the whole gas according to Maxwell-Boltzmann, Bose-Einstein, and Fermi-Dirac statistics, respectively.

For the case of Maxwell-Boltzmann (MB) statistics, the two particles are considered to be distinguishable. Let us denote them A and B. Furthermore, any number of particles can occupy the same quantum state.

Table. Two Particles Distributed Amongst three States According to Maxwell-Boltzmann Statistics.

1	2	3
AB	...	...
...	*AB*	...
...	...	*AB*
A	*B*	...
B	*A*	...
A	...	*B*
B	...	*A*
...	*A*	*B*
...	*B*	*A*

There are Clearly 9 Distinct States: For the case of Bose-Einstein (BE) statistics, the two particles are considered to be indistinguishable.

Let us denote them both as A. Furthermore, any number of particles can occupy the same quantum state. The possible different states of the gas are shown in Tab.

Table. Two Particles Distributed Amongst three States According to Bose-Einstein Statistics.

1	2	3
AA	...	...
...	*AA*	...
...	...	*AA*
A	*A*	...
A	...	*A*
...	*A*	*A*

There are clearly 6 distinct states: Finally, for the case of Fermi-Dirac (FD) statistics, the two particles are considered to be indistinguishable. Let us again denote them both as A. Furthermore, no more than one particle can occupy a given quantum state.

Table.Two Particles Distributed Amongst three States According to Fermi-Dirac Statistics.

1	2	3
A	A	...
A	...	A
...	A	A

There are clearly only 3 distinct states: It follows, from the above example, that Fermi-Dirac statistics are more restrictive (i.e., there are less possible states of the system) than Bose-Einstein statistics, which are, in turn, more restrictive than Maxwell-Boltzmann statistics. Let

$$\xi \equiv \frac{\text{probability that the two particles are found in the same state}}{\text{probability that the two particles are found in different state}}.$$

For the case under investigation,

$$\xi_{MB} = 1/2,$$
$$\xi_{BE} = 1,$$
$$\xi_{FD} = 0.$$

We conclude that in Bose-Einstein statistics there is a greater relative tendency for particles to cluster in the same state than in classical statistics.

On the other hand, in Fermi-Dirac statistics there is less tendency for particles to cluster in the same state than in classical statistics.

Formulation of the Statistical Problem

Consider a gas consisting of N identical non-interacting particles occupying volume V and in thermal equilibrium at temperature T. Let us label the possible quantum states of a single particle by τ (or s). Let the energy of a particle in state τ be denoted $\in_\tau$. Let the number of particles in state r be written n_τ. Finally, let us label the possible quantum states of the whole gas by R.

The particles are assumed to be non-interacting, so the total energy of the gas in state R, where there are n_τ particles in quantum state τ, etc., is simply

$$E_R = \sum_\tau n_\tau \in_\tau,$$

where the sum extends over all possible quantum states τ. Furthermore, since the total number of particles in the gas is known to be N, we must have

$$N = \sum_\tau n_\tau.$$

In order to calculate the thermodynamic properties of the gas (i.e., its internal energy or its entropy), it is necessary to calculate its partition function,

$$N = \sum_R e^{-\beta E_R} = \sum_R e^{-\beta(n_1 c_1 + n_2 c_2 + \cdots)}.$$

Here, the sum is over all possible states R of the whole gas: i.e., over all the various possible values of the numbers $n_1, n_2, \cdots$.

Now, $\exp[-\beta\,(n_1 \in_1 + n_2 \in_2 + \cdots)]$ is the relative probability of finding the gas in a particular state in which there are n_1 particles in state 1, n_2 particles in state 2, etc. Thus, the mean number of particles in quantum state s can be written

$$\bar{n}_s = \frac{\Sigma_R n_s \exp[-\beta(n_1 \in_1 + n_2 \in_2 + \cdots)]}{\Sigma_R \exp[-\beta(n_1 \in_1 + n_2 \in_2 + \cdots)]}.$$

A comparison of Eqs. yields the result

$$\bar{n}_s = \frac{1}{\beta}\frac{\partial \ln Z}{\partial \in_s}.$$

Here, $\beta \equiv 1/kT$.

FERMI-DIRAC STATISTICS

Let us, first of all, consider Fermi-Dirac statistics. According to Eq. the average number of particles in quantum state S can be written

$$\bar{n}_s = \frac{\sum_{n_s} n_s e^{-\beta n_s c_s} \sum^{(s)}_{n_1, n_2, \cdots} e^{-\beta(n_1 c_1 + n_2 c_2 + \cdots)}}{\sum_{n_s} e^{-\beta n_s c_s} \sum^{(s)}_{n_1, n_2, \cdots} e^{-\beta(n_1 c_1 + n_2 c_2 + \cdots)}}.$$

Here, we have rearranged the order of summation, using the multiplicative properties of the exponential function. Note that the first sums in the numerator and denominator only involve n_s, whereas the last sums omit the particular state s from consideration (this is indicated by the superscript s on the summation symbol). Of course, the sums in the above expression range over all values of the numbers $n_1, n_2, \ldots$ such that $n_\tau = 0$ and 1 for each τ, subject to the overall constraint that

$$\sum_{\tau} n_\tau = N.$$

Let us introduce the function

$$Z_s(N) = \sum_{n_1, n_2, \cdots}^{(s)} e^{-\beta(n_1 c_1 + n_2 c_2 + \cdots)},$$

which is defined as the partition function for N particles distributed over all quantum states, excluding state s, according to Fermi-Dirac statistics. By explicitly performing the sum over $n_s = 0$ and 1, the expression reduces to

$$\bar{n}_s = \frac{0 + e^{-\beta c_s} Z_s(N-1)}{Z_s(N) + e^{-\beta c_s} Z_s(N-1)},$$

which yields

$$\bar{n}_s = \frac{1}{[Z_s(N)/Z_s(N-1)]\, e^{-\beta c_s} + 1}$$

In order to make further progress, we must somehow relate $Z_s(N-1)$ to $Z_s(N)$. Suppose that $\Delta N \ll N$. It follows that $\ln Z_s(N - \Delta N)$ can be Taylor expanded to give

$$\ln Z_s(N - \Delta N) \simeq \ln Z_s(N) - \frac{\partial \ln Z_s}{\partial N} \Delta N = \ln Z_s(N) - \alpha_s \, \Delta N,$$

where

$$\alpha_s \equiv \frac{\partial \ln Z_s}{\partial N}.$$

As always, we Taylor expand the slowly varying function $\ln Z_s(N)$, rather than the rapidly varying function $Z_s(N)$, because the radius of convergence of the latter Taylor series is too small for the series to be of any practical use. Equation can be rearranged to give

$$Z_s(N - \Delta N) = Z_s(N)e^{-\alpha_s \Delta N}$$

Now, since $Z_s(N)$ is a sum over very many different quantum states, we would not expect the logarithm of this function to be sensitive to which particular state s is excluded from consideration. Let us, therefore, introduce the approximation that α_s is independent of s, so that we can write

$$\alpha_s \simeq \alpha$$

for all s. It follows that the derivative can be expressed approximately in terms of the derivative of the full partition function $Z(N)$ (in which the N particles are distributed over all quantum states). In fact,

$$\alpha \simeq \frac{\partial \ln Z}{\partial N}.$$

Making use of Eq. with $\Delta N = 1$, plus the approximation the expression reduces to

$$\bar{n}_s = \frac{1}{e^{\alpha+\beta c_s} + 1}.$$

This is called the Fermi-Dirac distribution. The parameter α is determined by the constraint that $\Sigma_r \bar{n}_r = N$: i.e.,

$$\sum_{\tau} \frac{1}{e^{\alpha+\beta c_r} + 1} = N.$$

Note that $\bar{n}_s \to 0$ if ϵ_s becomes sufficiently large. On the other hand, since the denominator in Eq. can never become less than unity, no matter how small ϵ_s becomes, it follows that $\bar{n}_s \leq 1$. Thus,

$$0 \leq \bar{n}_s \leq 1,$$

in accordance with the Pauli exclusion principle.

Equations can be integrated to give

$$\ln Z = \alpha N + \sum_{\tau} \ln(1 + e^{-\alpha-\beta c_\tau}),$$

where use has been made of Eq.

PHOTON STATISTICS

Up to now, we have assumed that the number of particles N contained in a given system is a fixed number. This is a reasonable assumption if the particles possess non-zero mass, since we are not generally considering relativistic systems in this course.

However, this assumption breaks down for the case of photons, which are zero-mass bosons. In fact, photons enclosed in a container of volume V, maintained at temperature T, can readily be absorbed or emitted by the walls. Thus, for the special case of a gas of photons there is no requirement which limits the total number of particles.

It follows, from the above discussion, that photons obey a simplified form of Bose-Einstein statistics in which there is an unspecified total number of particles. This type of statistics is called photon statistics.

Consider the expression. For the case of photons, the numbers $n_1, n_2 \ldots$ assume all values $n_\tau = 0, 1, 2, \ldots$ for each τ, without any further restriction. It follows that the sums $\Sigma^{(s)}$in the numerator and denominator are identical and, therefore, cancel. Hence, Eq. reduces to

$$\bar{n}_s = \frac{\Sigma_{n_s} n_s e^{-\beta n_s c_s}}{\Sigma_{n_s} e^{-\beta n_s c_s}}.$$

However, the above expression can be rewritten

$$\bar{n}_s = -\frac{1}{\beta}\frac{\partial}{\partial \epsilon_s}\left(\ln \sum_{n_s} e^{-\beta n_s c_s}\right).$$

Now, the sum on the right-hand side of the above equation is an infinite geometric series, which can easily be evaluated. In fact,

$$\sum_{n_s=0}^{\infty} e^{-\beta n_s c_s} = 1 + e^{-\beta c_s} + e^{-2\beta c_s} + \cdots = \frac{1}{1 - e^{-\beta c_s}}.$$

Thus, Eq. gives

$$\bar{n}_s = \frac{1}{\beta}\frac{\partial}{\partial \epsilon_s}\ln(1 - e^{-\beta c_s}) = \frac{e^{-\beta c_s}}{1 - e^{-\beta c_s}},$$

or

$$\bar{n}_s = \frac{1}{e^{\beta c_s} - 1}.$$

This is known as the Planck distribution, after the German physicist Max Planck who first proposed it in 1900 on purely empirical grounds.

Equation can be integrated to give

$$\ln Z = -\sum_{\tau} \ln(1 - e^{-\beta c_r}),$$

where use has been made of Eq.

BOSE-EINSTEIN STATISTICS

Let us now consider Bose-Einstein statistics. The particles in the system are assumed to be massive, so the total number of particles N is a fixed number.

Consider the expression. For the case of massive bosons, the numbers $n_1, n_2, \ldots$ assume all values $n_\tau = 0, 1, 2, \ldots$ for each r, subject to the constraint that $\Sigma_\tau n_\tau = N$. Performing explicitly the sum over n_s, this expression reduces to

$$\bar{n}_s = \frac{0 + e^{-\beta c_s} Z_s(N-1) + 2e^{-2\beta c_s} Z_s(N-2) + \cdots}{Z_s(N) + e^{-\beta c_s} Z_s(N-1) + e^{-2\beta c_s} Z_s(N-2) + \cdots},$$

where $Z_s(N)$is the partition function for N particles distributed over all quantum states, excluding state s, according to Bose-Einstein statistics. Using Eq., and the approximation the above equation reduces to

$$\bar{n}_s = \frac{\Sigma_s n_s e^{-n_s(\alpha + \beta c_s)}}{\Sigma_s e^{-n_s(\alpha + \beta c_s)}}.$$

Note that this expression is identical to except that $\beta \in_s$ is replaced by $\alpha + \beta \in_s$. Hence, an analogous calculation to that outlined in the previous subsection yields

$$\bar{n}_s = \frac{1}{e^{\alpha + \beta c_s} - 1}.$$

This is called the Bose-Einstein distribution. Note that $\bar{n}_s$ can become very large in this distribution. The parameter α is again determined by the constraint on the total number of particles: i.e.,

$$\sum_{\tau} \frac{1}{e^{\alpha+\beta c_{\tau}} - 1} = N.$$

Equations can be integrated to give

$$\ln Z = \alpha N - \sum_{\tau} \ln(1 - e^{\alpha-\beta c_{\tau}}),$$

where use has been made of Eq.

Note that photon statistics correspond to the special case of Bose-Einstein statistics in which the parameter α takes the value zero, and the constraint does not apply.

MAXWELL-BOLTZMANN STATISTICS

For the purpose of comparison, it is instructive to consider the purely classical case of Maxwell-Boltzmann statistics. The partition function is written

$$Z = \sum_{R} e^{-\beta(a_1 c_1 + n_2 c_2 + \cdots)},$$

where the sum is over all distinct states R of the gas, and the particles are treated as distinguishable. For given values of $n_1, n_2 \ldots$ there are

$$\frac{N!}{n_1!n_2!\cdots}$$

possible ways in which N distinguishable particles can be put into individual quantum states such that there are n_1 particles in state 1, n_2 particles in state 2, etc. Each of these possible arrangements corresponds to a distinct state for the whole gas. Hence, Eq. can be written

$$Z = \sum_{n_1!n_2!\cdots} \frac{N!}{n_1!n_2!\cdots} e^{-\beta(n_1 c_1 + n_2 c_2 + \cdots)},$$

where the sum is over all values of $n_{\tau} = 0, 1, 2, \ldots$ for each r, subject to the constraint that

$$\sum_{\tau} n_{\tau} = N.$$

Now, Eq. can be written

$$Z = \sum_{n_1, n_2 \ldots} \frac{N!}{n_1! n_2!} (e^{-\beta c_1})^{n_1} (e^{-\beta c_2})^{n_2} \ldots,$$

which, by virtue of Eq. is just the result of expanding a polynomial. In fact,

$$Z = (e^{-\beta c_1} + e^{-\beta c_2} + \cdots)^N,$$

or

$$\ln Z = N \ln\left(\sum_{\tau} e^{-\beta c_\tau}\right).$$

Note that the argument of the logarithm is simply the partition function for a single particle.

Equations can be combined to give

$$\overline{n}_s = N \frac{e^{-\beta c_s}}{\Sigma_r e^{-\beta c_r}}.$$

This is known as the Maxwell-Boltzmann distribution. It is, of course, just the result obtained by applying the Boltzmann distribution to a single particle

QUANTUM STATISTICS IN THE CLASSICAL LIMIT

The preceding analysis regarding the quantum statistics of ideal gases is summarized in the following statements. The mean number of particles occupying quantum state s is given by

$$\overline{n}_s = \frac{1}{e^{\alpha + \beta c_s} \pm 1},$$

where the upper sign corresponds to Fermi-Dirac statistics and the lower sign corresponds to Bose-Einstein statistics. The parameter α is determined via

$$\sum_{\tau} \overline{n}_s = \sum_{\tau} \frac{1}{e^{\alpha + \beta c_s} \pm 1} = N.$$

Finally, the partition function of the gas is given by

$$\ln Z = \alpha N \pm \sum_{\tau} \ln(1 \pm e^{-\alpha - \beta c_r}).$$

Let us investigate the magnitude of α in some important limiting cases. Consider, first of all, the case of a gas at a given temperature when its concentration is made sufficiently low: i.e., when N is made sufficiently small. The relation can only be satisfied if each term in the sum over states is made sufficiently small; i.e., if $\bar{n}_\tau \ll 1$ or $\exp(\alpha + \beta \in_\tau) \gg 1$ for all states τ.

Consider, next, the case of a gas made up of a fixed number of particles when its temperature is made sufficiently large: i.e., when β is made sufficiently small. In the sum in Eq. the terms of appreciable magnitude are those for which $\beta \in_\tau \ll \alpha$. Thus, it follows that as $\beta \to 0$ an increasing number of terms with large values of $\in_\tau$ contribute substantially to this sum. In order to prevent the sum from exceeding N, the parameter α must become large enough that each term is made sufficiently small: i.e., it is again necessary that $\bar{n}_\tau \ll 1$ or $\exp(\alpha + \beta \in_\tau) \gg$ for all states τ.

The above discussion suggests that if the concentration of an ideal gas is made sufficiently low, or the temperature is made sufficiently high, then α must become so large that

$$e^{\alpha + \beta c_\tau} >> 1$$

for all τ. Equivalently, this means that the number of particles occupying each quantum state must become so small that

$$\bar{n}_\tau << 1$$

for all τ. It is conventional to refer to the limit of sufficiently low concentration, or sufficiently high temperature, in which Eqs. are satisfied, as the classical limit.

According to Eqs. both the Fermi-Dirac and Bose-Einstein distributions reduce to

$$\bar{n}_s = e^{-\alpha - \beta c_s}$$

in the classical limit, whereas the constraint yields

$$\sum_\tau e^{-\alpha - \beta c_\tau} = N.$$

The above expressions can be combined to give

$$\bar{n}_s = N \frac{e^{-\beta c_s}}{\sum_\tau e^{-\beta c_\tau}}.$$

It follows that in the classical limit of sufficiently low density, or sufficiently high temperature, the quantum distribution functions, whether Fermi-Dirac or Bose-Einstein, reduce to the Maxwell-Boltzmann distribution. It is easily demonstrated that the physical criterion for the validity of the classical approximation is that the mean separation between particles should be much greater than their mean de Broglie wavelengths.

Let us now consider the behaviour of the partition function in the classical limit. We can expand the logarithm to give

$$\ln Z = \alpha\, N \pm \sum_{\tau} \left(\pm\, e^{-\alpha - \beta c_\tau} \right) = \alpha N + N.$$

However, according to Eq.

$$\alpha = -\ln N + \ln\left(\sum_{\tau} e^{-\beta c_\tau} \right).$$

It follows that

$$\ln Z = -N \ln N + N + N \ln\left(\sum_{\tau} e^{-\beta c_\tau} \right).$$

Note that this does not equal the partition function Z_{MB} computed in Eq. from Maxwell-Boltzmann statistics: i.e.,

$$\ln Z_{MB} = N \ln\left(\sum_{\tau} e^{-\beta c_\tau} \right).$$

In fact,

$$\ln Z = \ln Z_{MB} - \ln N!,$$

or

$$Z = \frac{Z_{MB}}{N!},$$

where use has been made of Stirling's approximation ($N! \simeq N \ln N - N$), since N is large. Here, the factor $N!$ simply corresponds to the number of different permutations of the N particles: permutations which are physically meaningless when the particles are identical. Recall, that we had to introduce precisely this factor, in an ad hoc fashion in order to avoid the non-physical consequences of the Gibb's paradox.

Clearly, there is no Gibb's paradox when an ideal gas is treated properly via quantum mechanics.

In the classical limit, a full quantum mechanical analysis of an ideal gas reproduces the results obtained , except that the arbitrary parameter h_0 is replaced by Planck's constant $h = 6.61 \times 10^{-34}$ Js.

A gas in the classical limit, where the typical de Broglie wavelength of the constituent particles is much smaller than the typical inter-particle spacing, is said to be non-degenerate. In the opposite limit, where the concentration and temperature are such that the typical de Broglie wavelength becomes comparable with the typical inter-particle spacing, and the actual Fermi-Dirac or Bose-Einstein distributions must be employed, the gas is said to be degenerate.

THE PLANCK RADIATION LAW

Let us now consider the application of statistical thermodynamics to electromagnetic radiation. According to Maxwell's theory, an electromagnetic wave is a coupled self-sustaining oscillation of electric and magnetic fields which propagates though a vacuum at the speed of light, $c = 3 \times 10^8$ ms^{-1}. The electric component of the wave can be written

$$E = E_0 \exp[i(k.r - \omega t],$$

where E_0 is a constant, k is the wave-vector which determines the wavelength and direction of propagation of the wave, and ω is the frequency. The dispersion relation

$$\omega = kc$$

ensures that the wave propagates at the speed of light. Note that this dispersion relation is very similar to that of sound waves in solids. Electromagnetic waves always propagate in the direction perpendicular to the coupled electric and magnetic fields (i.e., electromagnetic waves are transverse waves).

This means that $k.E_0 = 0$. Thus, once k is specified, there are only two possible independent directions for the electric field. These correspond to the two independent polarizations of electromagnetic waves.

Consider an enclosure whose walls are maintained at fixed temperature T. What is the nature of the steady-state electromagnetic radiation inside the enclosure? Suppose that the enclosure is a parallelepiped with sides of lengths L_x, L_y, and L_z.

Alternatively, suppose that the radiation field inside the enclosure is periodic in the x–, y–, and z–directions, with periodicity lengths L_x, L_y, and L_z, respectively. As long as the smallest of these lengths, L, say, is much greater than the longest wavelength of interest in the problem, $\lambda = 2\pi/k$, then these assumptions should not significantly affect the nature of the radiation inside the enclosure.

We find, just as in our earlier discussion of sound waves, that the periodicity constraints ensure that there are only a discrete set of allowed wave-vectors (i.e., a discrete set of allowed modes of oscillation of the electromagnetic field inside the enclosure). Let $\rho(k)\, d^3k$ be the number of allowed modes per unit volume with wave-vectors in the range k to $k + dk$. We know, by analogy with Eq. that

$$\rho(k)d^3k = \frac{d^3k}{(2\pi)^3}.$$

The number of modes per unit volume for which the magnitude of the wave-vector lies in the range k to $k + dk$ is just the density of modes, $\rho(k)$, multiplied by the "volume" in k-space of the spherical shell lying between radii k and $k + dk$. Thus,

$$\rho_k(k)dk = \frac{4\pi k^2 dk}{(2\pi)^3} = \frac{k^2}{2\pi^2}dk.$$

Finally, the number of modes per unit volume whose frequencies lie between ω and $\omega + d\omega$ is, by Eq.),

$$\sigma(\omega)d\omega = 2\frac{\omega^2 dk}{2\pi^2 c^3}d\omega.$$

Here, the additional factor 2 is to take account of the two independent polarizations of the electromagnetic field for a given wave-vector k.

Let us consider the situation classically. By analogy with sound waves, we can treat each allowable mode of oscillation of the electromagnetic field as an independent harmonic oscillator. According to the equipartition theorem, each mode possesses a mean energy kT in thermal equilibrium at temperature T. In fact, (1/2) kT resides with the oscillating electric field, and another (1/2)kT with the oscillating magnetic field. Thus, the classical energy density of electromagnetic radiation (i.e., the energy per unit volume associated with modes whose frequencies lie in the range ω to $\omega + d\omega$) is

$$\overline{u}(\omega)d\omega = kT\sigma(\omega)d\omega \frac{kT}{\pi^2 c^3}\omega^2 d\omega.$$

This result is known as the Rayleigh-Jeans radiation law, after Lord Rayleigh and James Jeans who first proposed it in the late nineteenth century.

According to Debye theory, the energy density of sound waves in a solid is analogous to the Rayleigh-Jeans law, with one very important difference. In Debye theory there is a cut-off frequency (the Debye frequency) above which no modes exist. This cut-off comes about because of the discrete nature of solids (i.e., because solids are made up of atoms instead of being continuous). It is, of course, impossible to have sound waves whose wavelengths are much less than the inter-atomic spacing. On the other hand, electromagnetic waves propagate through a vacuum, which possesses no discrete structure. It follows that there is no cut-off frequency for electromagnetic waves, and so the Rayleigh-Jeans law holds for all frequencies. This immediately poses a severe problem. The total classical energy density of electromagnetic radiation is given by

$$U = \int_0^\infty \overline{u}(\omega)d\omega = \frac{kT}{\pi^2 C^3}\int_0^\infty \omega^2 d\omega.$$

This is an integral which obviously does not converge. Thus, according to classical physics, the total energy density of electromagnetic radiation inside an enclosed cavity is infinite! This is clearly an absurd result, and was recognized as such in the latter half of the nineteenth century. In fact, this prediction is known as the ultra-violet catastrophe, because the Rayleigh-Jeans law usually starts to diverge badly from experimental observations (by over-estimating the amount of radiation) in the ultra-violet region of the spectrum.

So, how do we obtain a sensible answer? Well, as usual, quantum mechanics comes to our rescue. According to quantum mechanics, each allowable mode of oscillation of the electromagnetic field corresponds to a photon state with energy and momentum

$$\in = \hbar\omega,$$

$$P = \hbar k,$$

respectively. Incidentally, it follows from Eq. that

$$\in = pc,$$

which implies that photons are massless particles which move at the speed of light. According to the Planck distribution, the mean number of photons occupying a photon state of frequency ω is

$$n(\omega) = \frac{1}{e^{\beta\hbar\omega} - 1}.$$

Hence, the mean energy of such a state is given by

$$\bar{\in}(\omega) = \hbar\omega\, n(\omega) = \frac{\hbar\omega}{e^{\beta\hbar\omega} - 1}.$$

Note that low frequency states (i.e., $\hbar\omega \ll kT$) behave classically: i.e.,

$$\bar{\in} \simeq kT.$$

On the other hand, high frequency states (i.e., $\hbar\omega >> kT$) are completely "frozen out": i.e.,

$$\bar{\in} << kT.$$

The reason for this is simply that it is very difficult for a thermal fluctuation to create a photon with an energy greatly in excess of *kT*, since *kT* is the characteristic energy associated with such fluctuations.

According to the above discussion, the true energy density of electromagnetic radiation inside an enclosed cavity is written

$$\bar{u}d\omega = \in(\omega)\sigma(\omega)d\omega,$$

giving

$$\bar{u}d\omega = \frac{\hbar}{\pi^2 C^3}\frac{\omega^3 d\omega}{\exp(\beta\hbar\omega)-1}.$$

This is famous result is known as the Planck radiation law. The Planck law approximates to the classical Rayleigh-Jeans law for $\hbar\omega << kT$, peaks at about $\hbar\omega \simeq 3kT$, and falls off exponentially for $\hbar\omega >> kT$. The exponential fall off at high frequencies ensures that the total energy density remains finite.

BLACK-BODY RADIATION

Suppose that we were to make a small hole in the wall of our enclosure, and observe the emitted radiation. A small hole is the best approximation in Physics to a black-body, which is defined as an object which absorbs, and, therefore, emits, radiation perfectly at all wavelengths. What is the power radiated by the hole? Well, the power density inside the enclosure can be written

$$u(\omega)d\omega = \hbar\omega\, n(\omega)d\omega,$$

where $n(\omega)$ is the mean number of photons per unit volume whose frequencies lie in the range ω to $\omega + d\omega$. The radiation field inside the enclosure is isotropic (we are assuming that the hole is sufficiently small that it does not distort the field). It follows that the mean number of photons per unit volume whose frequencies lie in the specified range, and whose directions of propagation make an angle in the range θ to $\theta + d\theta$ with the normal to the hole, is

$$u(\omega,\theta)d\omega\, d\theta = \frac{1}{2}n(\omega)\, d\omega \sin\theta\, d\theta,$$

where $\sin\theta$ is proportional to the solid angle in the specified range of directions, and

$$\int_0^{\pi} n(\omega,\theta)d\omega\, d\theta = n(\omega)\, d\omega.$$

Photons travel at the velocity of light, so the power per unit area escaping from the hole in the frequency range ω to $\omega + d\omega$ is

$$P(\omega)d\omega \int_0^{\pi/2} c\cos\theta\hbar\omega\, n(\omega,\theta)\, d\omega\, d\theta,$$

where $c \cos\theta$ is the component of the photon velocity in the direction of the hole. This gives

$$P(\omega)\,d\omega = c\bar{u}(\omega)\,d\omega\frac{1}{2}\int_0^{\pi/2}\cos\theta\sin\theta\,d\theta = \frac{c}{4}\bar{u}(\omega)\,d\omega,$$

so

$$P(\omega)\,d\omega = \frac{\hbar}{4\pi^2c^2}\frac{\omega^3 d\omega}{\exp(\beta\hbar\omega)-1}$$

is the power per unit area radiated by a black-body in the frequency range ω to $\omega + d\omega$.

A black-body is very much an idealization. The power spectra of real radiating bodies can deviate quite substantially from black-body spectra. Nevertheless, we can make some useful predictions using this model.

The black-body power spectrum peaks when $\hbar\omega \simeq 3kT$. This means that the peak radiation frequency scales linearly with the temperature of the body. In other words, hot bodies tend to radiate at higher frequencies than cold bodies.

This result (in particular, the linear scaling) is known as Wien's displacement law.

It allows us to estimate the surface temperatures of stars from their colours (surprisingly enough, stars are fairly good black-bodies).

Some stellar temperatures determined by this method (in fact, the whole emission spectrum is fitted to a black-body spectrum). It can be seen that the apparent colours (which correspond quite well to the colours of the peak radiation) scan the whole visible spectrum, from red to blue, as the stellar surface temperatures gradually rise.

Table. Physical Properties of Some Well-known Stars

Name	*Constellation*	*Spectral Type*	*Surf. Temp. (°K)*	*Colour*
Antares	Scorpio	M	3300	Very Red
Aldebaran	Taurus	K	3800	Reddish
Sun		G	5770	Yellow
Procyon	Canis Minor	F	6570	Yellowish
Sirius	Canis Major	A	9250	White
Rigel	Orion	B	11,200	Bluish White

Probably the most famous black-body spectrum is cosmological in origin. Just after the "big bang" the Universe was essentially a "fireball," with the energy associated with radiation completely dominating that associated with matter. The early Universe was also pretty well described by equilibrium statistical thermodynamics, which means that the radiation had a black-body spectrum. As the Universe expanded, the radiation was gradually Doppler shifted to ever larger wavelengths (in other words, the radiation did work against the expansion of the Universe, and, thereby, lost energy), but its spectrum remained invariant.

Nowadays, this primordial radiation is detectable as a faint microwave background which pervades the whole universe. The microwave background was discovered accidentally by Penzias and Wilson in 1961. Until recently, it was difficult to measure the full spectrum with any degree of precision, because of strong microwave absorption and scattering by the Earth's atmosphere.

However, all of this changed when the COBE satellite was launched in 1989. It took precisely nine minutes to measure the perfect black-body spectrum reproduced. This data can be fitted to a black-body curve of characteristic temperature 2.735°K. In a very real sense, this can be regarded as the "temperature of the Universe."

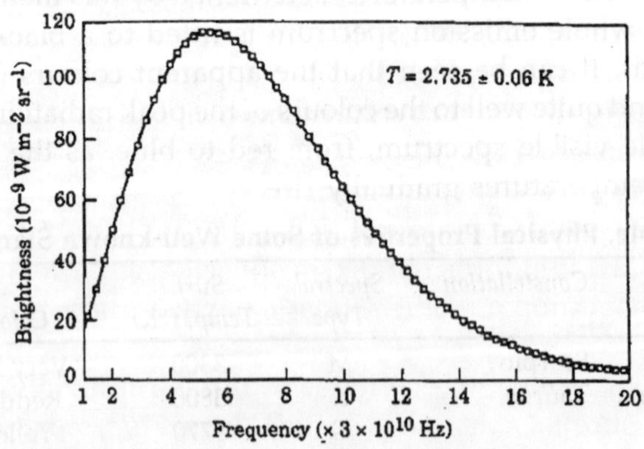

Fig. Cosmic Background Radiation Spectrum Measured by the Far Infrared Absolute Spectrometer (FIRAS) aboard the Cosmic Background Explorer Satellite (COBE).

THE STEFAN-BOLTZMANN LAW

The total power radiated per unit area by a black-body at all frequencies is given by

$$P_{\text{tot}}(T) = \int_0^\infty P(\omega)\, d\omega = \frac{\hbar}{4\pi^2 c^2} \int_0^\infty \frac{\omega^3 d\omega}{\exp(\hbar\omega / kT) - 1},$$

or

$$P_{\text{tot}}(T) = \frac{k^4}{4\pi^2 c^2 \hbar^3} \int_0^\infty \frac{\eta^3 d\eta}{\exp\eta - 1},$$

where $\eta = \hbar\omega / kT$. The above integral can easily be looked up in standard mathematical tables. In fact,

$$\int_0^\infty \frac{\eta^3 d\eta}{\exp\eta - 1} = \frac{\pi^4}{15}.$$

Thus, the total power radiated per unit area by a black-body is

$$P_{\text{tot}}(T) = \frac{\pi^2}{60} \frac{k^4}{c^2 \hbar^3} T^4 = \sigma T^4.$$

This T^4 dependence of the radiated power is called the Stefan-Boltzmann law, after Josef Stefan, who first obtained it experimentally, and Ludwig Boltzmann, who first derived it theoretically.

The parameter

$$\sigma = \frac{\pi^2}{60} \frac{k^4}{c^2 \hbar^3} = 5.67 \times 10^{-8}\ \text{W m}^{-2}\ \text{K}^{-4},$$

is called the Stefan-Boltzmann constant.

We can use the Stefan-Boltzmann law to estimate the temperature of the Earth from first principles. The Sun is a ball of glowing gas of radius $R_\odot \simeq 7 \times 10^5$ km and surface temperature $T_\odot \simeq 5770°$ K. Its luminosity is

$$L_\odot = 4\pi R_\odot^2 \sigma T_\odot^4,$$

according to the Stefan-Boltzmann law. The Earth is a globe of radius $R_\oplus \sim 6000$ km located an average distance

$\tau_\oplus \simeq 1.5\times10^8$ km from the Sun. The Earth intercepts an amount of energy

$$P_\oplus = L_\odot \frac{\pi R_\oplus^2 / \tau_\oplus^2}{4\pi}$$

per second from the Sun's radiative output: i.e., the power output of the Sun reduced by the ratio of the solid angle subtended by the Earth at the Sun to the total solid angle 4π. The Earth absorbs this energy, and then re-radiates it at longer wavelengths. The luminosity of the Earth is

$$L_\oplus = 4\pi R_\oplus^2 \sigma T_\oplus^4,$$

according to the Stefan-Boltzmann law, where $T_\oplus$ is the average temperature of the Earth's surface. Here, we are ignoring any surface temperature variations between polar and equatorial regions, or between day and night. In steady-state, the luminosity of the Earth must balance the radiative power input from the Sun, so equating $L_\oplus$ and $P_\oplus$ we arrive at

$$T_\oplus = \left(\frac{R_\odot}{2\tau_\oplus}\right)^{1/2} T_\odot.$$

Remarkably, the ratio of the Earth's surface temperature to that of the Sun depends only on the Earth-Sun distance and the solar radius. The above expression yields $T_\oplus \sim 279°$ K or 6° C (or 43°F). This is slightly on the cold side, by a few degrees, because of the greenhouse action of the Earth's atmosphere, which was neglected in our calculation. Nevertheless, it is quite encouraging that such a crude calculation comes so close to the correct answer.

CONDUCTION ELECTRONS IN A METAL

The conduction electrons in a metal are non-localized (i.e., they are not tied to any particular atoms). In conventional metals, each atom contributes a single such electron. To a first approximation, it is possible to neglect the mutual interaction of the conduction electrons, since this interaction is largely shielded out by the stationary atoms. The conduction electrons can, therefore, be treated as an ideal gas. However, the concentration of such electrons in a metal far exceeds the concentration of particles in a conventional gas. It is, therefore, not surprising that conduction electrons cannot normally be analyzed using classical statistics: in fact, they are subject to Fermi-Dirac statistics (since electrons are fermions).

Recall, that the mean number of particles occupying state s (energy $\in_s$) is given by

$$\bar{n}_s = \frac{1}{e^{\beta(c_s - \mu)} + 1},$$

according to the Fermi-Dirac distribution. Here,

$$\mu \equiv -kT\,\alpha$$

is termed the Fermi energy of the system. This energy is determined by the condition that

$$\sum_\tau \bar{n}_\tau = \sum_\tau \frac{1}{e^{\beta(c_\tau - \mu) + 1}} = N,$$

where N is the total number of particles contained in the volume V. It is clear, from the above equation, that the Fermi energy μ is generally a function of the temperature T.

Let us investigate the behaviour of the Fermi function

$$F(\in) = \frac{1}{e^{b(c-\mu)} + 1}$$

as $\in$ varies. Here, the energy is measured from its lowest possible value $\in = 0$. If the Fermi energy μ is such that $\beta\,\mu << 1$ then $\beta(\in - \mu) >> 1$, and F reduces to the Maxwell-Boltzmann distribution. However, for the case of conduction electrons in a metal we are interested in the opposite limit, where

$$\beta\mu \equiv \frac{\mu}{kT} >> 1.$$

In this limit, if $\in << \mu$ then $\beta(\in - \mu) << 1$, so that $F(\in) = 1$. On the other hand, if $\in >> \mu$ then $\beta(\in - \mu) >> 1$, so that $F(\in) = \exp[-\beta(\in - \mu)]$ falls off exponentially with increasing $\in$, just like a classical Boltzmann distribution. Note that $F = 1/2$ when $\in = \mu$. The transition region in which F goes from a value close to unity to a value close to zero corresponds to an energy interval of order kT, centred on $\in = \mu$.

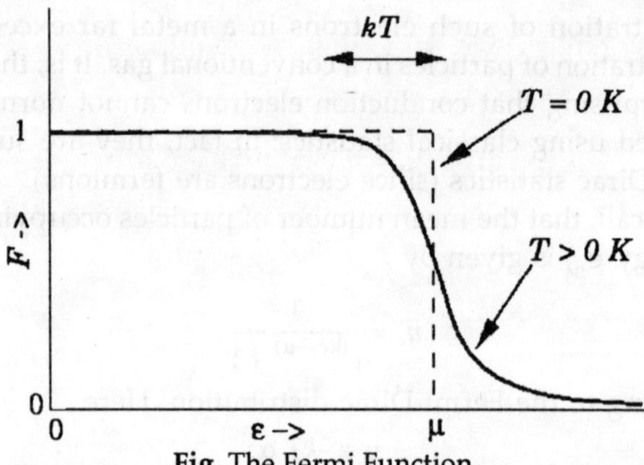

Fig. The Fermi Function.

In the limit as $T \to 0$, the transition region becomes infinitesimally narrow. In this case, $F = 1$ for $\in < \mu$ and $F = 0$ for $\in > \mu$. This is an obvious result, since when T = 0the conduction electrons attain their lowest energy, or ground-state, configuration. Since the Pauli exclusion principle requires that there be no more than one electron per single-particle quantum state, the lowest energy configuration is obtained by piling electrons into the lowest available unoccupied states until all of the electrons are used up. Thus, the last electron added to the pile has quite a considerable energy, $\in = \mu$, since all of the lower energy states are already occupied. Clearly, the exclusion principle implies that a Fermi-Dirac gas possesses a large mean energy, even at absolute zero.

Let us calculate the Fermi energy $\mu = \mu_0$ of a Fermi-Dirac gas at $T = 0$. The energy of each particle is related to its momentum $p = \hbar k$ via

$$\in = \frac{p^2}{2m} = \frac{\hbar^2}{2m},$$

where k is the de Broglie wave-vector. At $T = 0$ all quantum states whose energy is less than the Fermi energy μ_0 are filled. The Fermi energy corresponds to a Fermi momentum $P_F = \hbar$ $k_F$which is such that

$$\mu_0 = \frac{p_F^2}{2m} = \frac{\hbar^2 k_F^2}{2m}.$$

Thus, at $T = 0$ all quantum states with $k < k_F$ are filled, and all those with $k > k_F$ are empty.

Now, we know, by analogy with Eq. that there are $(2\pi)^{-3}$ V allowable translational states per unit volume of k-space. The volume of the sphere of radius k_F in k-space is $(4/3)\pi k_F^2$ .It follows that the Fermi sphere of radius k_F contains $(4/3) \pi k_F^2 (2\pi)^{-3} V$ translational states.

The number of quantum states inside the sphere is twice this, because electrons possess two possible spin states for every possible translational state. Since the total number of occupied states (i.e., the total number of quantum states inside the Fermi sphere) must equal the total number of particles in the gas, it follows that

$$2\frac{V}{(2\pi)^3}\left(\frac{4}{3}\pi k_F^3\right) = N.$$

The above expression can be rearranged to give

$$k_F = \left(3\pi^2 \frac{N}{V}\right)^{1/3}.$$

Hence,

which implies that the de Broglie wavelength λ_F corresponding to the Fermi energy is of order the mean separation between particles $(V/N)^{1/3}$. All quantum states with de Broglie wavelengths $l \equiv 2\pi/k > \lambda_F$ are occupied at $T = 0$, whereas all those with $\lambda < \lambda_F$ are empty.

According to Eq. the Fermi energy at $T = 0$ takes the form

$$\mu_0 = \frac{\hbar^2}{2m}\left(3\pi^2 \frac{N}{V}\right)^{1/3}.$$

It is easily demonstrated that $\mu_0 >> kT$ for conventional metals at room temperature.

The majority of the conduction electrons in a metal occupy a band of completely filled states with energies far below the Fermi energy. In many cases, such electrons have very little effect on the macroscopic properties of the metal. Consider, for example, the contribution of the conduction electrons to the specific heat of the metal. The heat capacity C_V at constant volume of these electrons can be calculated from a knowledge of their mean energy $\bar{E}(T)$ as a function of T: i.e.,

$$C_V = \left(\frac{\partial \bar{E}}{\partial T}\right)_V.$$

If the electrons obeyed classical Maxwell-Boltzmann statistics, so that $F \propto \exp(-\beta \in)$ for all electrons, then the equipartition theorem would give

$$\bar{E} = \frac{3}{2} NkT,$$

$$C_V = \frac{3}{2} Nk.$$

However, the actual situation, in which F has the form is very different. A small change in T does not affect the mean energies of the majority of the electrons, with $\in << \mu$, since these electrons lie in states which are completely filled, and remain so when the temperature is changed. It follows that these electrons contribute nothing whatsoever to the heat capacity.

On the other hand, the relatively small number of electrons N_{eff} in the energy range of order kT, centred on the Fermi energy, in which $F$is significantly different from 0 and 1, do contribute to the specific heat. In the tail end of this region $F \propto \exp(-\beta \in)$, so the distribution reverts to a Maxwell-Boltzmann distribution. Hence, from Eq. we expect each electron in this region to contribute roughly an amount $(3/2)k$to the heat capacity. Hence, the heat capacity can be written

$$C_V \simeq \frac{3}{2} N_{eff} k.$$

However, since only a fraction kT/μ of the total conduction electrons lie in the tail region of the Fermi-Dirac distribution, we expect

$$N_{eff} \simeq \frac{kT}{\mu} N.$$

It follows that

$$C_V \simeq \frac{3}{2} N k \frac{kT}{\mu}.$$

Since $kT << \mu$ in conventional metals, the molar specific heat of the conduction electrons is clearly very much less than the classical value $(3/2)R$. This accounts for the fact that the molar specific heat capacities of metals at room temperature are about the same as those of insulators. Before the advent of quantum mechanics, the classical theory predicted incorrectly that the presence of conduction electrons should raise the heat capacities of metals by 50 percent [i.e., $(3/2)R$] compared to those of insulators.

Note that the specific heat is not temperature independent. In fact, using the superscript e to denote the electronic specific heat, the molar specific heat can be written

$$C_V^{(e)} = \gamma T,$$

where γ is a (positive) constant of proportionality. At room temperature $C_V^{(e)}$ is completely masked by the much larger specific heat $C_V^{(L)}$ due to lattice vibrations. However, at very low temperatures $C_V^{(L)} = AT^3$, where A is a (positive) constant of proportionality. Clearly, at low temperatures $C_V^{(L)} = AT^3$ approaches zero far more rapidly that the electronic specific heat, as T is reduced. Hence, it should be possible to measure the electronic contribution to the molar specific heat at low temperatures.

The total molar specific heat of a metal at low temperatures takes the form

$$c_V = C_V^{(e)} + c_V^{(L)} = \gamma T + AT^3.$$

Hence,

$$\frac{c_V}{T} = \gamma + AT^2.$$

If follows that a plot of c_V/T versus T^2 should yield a straight line whose intercept on the vertical axis gives the coefficient γ. The fact that a good straight line is obtained verifies that the temperature dependence of the heat capacity predicted by Eq. is indeed correct.

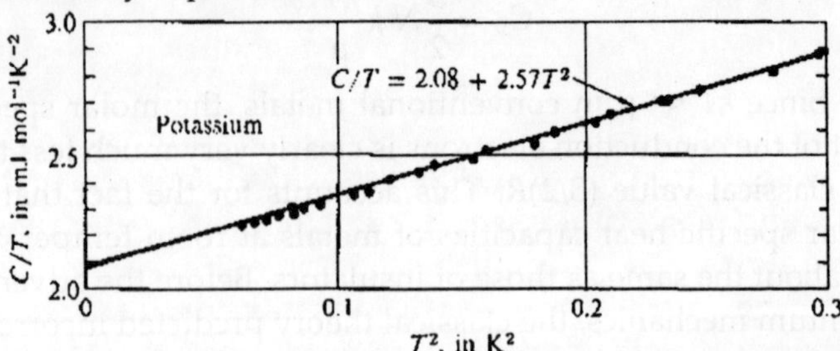

Fig. The Low Temperature Heat Capacity of Potassium, Plotted as C_V/T versus T^2. From C. Kittel, and H. Kroemer, Themal Physics.

White-Dwarf Stars

A main-sequence hydrogen-burning star, such as the Sun, is maintained in equilibrium via the balance of the gravitational attraction tending to make it collapse, and the thermal pressure tending to make it expand. Of course, the thermal energy of the star is generated by nuclear reactions occurring deep inside its core.

Eventually, however, the star will run out of burnable fuel, and, therefore, start to collapse, as it radiates away its remaining thermal energy. What is the ultimate fate of such a star?

A burnt-out star is basically a gas of electrons and ions. As the star collapses, its density increases, so the mean separation between its constituent particles decreases. Eventually, the mean separation becomes of order the de Broglie wavelength of the electrons, and the electron gas becomes degenerate. Note, that the de Broglie wavelength of the ions is much smaller than that of the electrons, so the ion gas remains non-degenerate.

Now, even at zero temperature, a degenerate electron gas exerts a substantial pressure, because the Pauli exclusion principle prevents the mean electron separation from becoming significantly smaller than the typical de Broglie wavelength. Thus, it is possible for a burnt-out star to maintain itself against complete collapse under gravity via the degeneracy pressure of its constituent electrons. Such stars are termed white-dwarfs. Let us investigate the physics of white-dwarfs in more detail.

The total energy of a white-dwarf star can be written

$$E = K + U,$$

where K is the total kinetic energy of the degenerate electrons (the kinetic energy of the ion is negligible) and U is the gravitational potential energy. Let us assume, for the sake of simplicity, that the density of the star is uniform. In this case, the gravitational potential energy takes the form

$$U = -\frac{3}{5}\frac{G\,M^2}{R},$$

where G is the gravitational constant, M is the stellar mass, and R is the stellar radius.

Let us assume that the electron gas is highly degenerate, which is equivalent to taking the limit $T \to 0$. In this case, that the Fermi momentum can be written

$$p_F = \Lambda\left(\frac{N}{V}\right)^{1/3},$$

where

$$\Lambda = \left(3p^2\right)^{1/3} \hbar.$$

Here,

$$V = \frac{4\pi}{3}R^3$$

is the stellar volume, and N is the total number of electrons contained in the star. Furthermore, the number of electron states contained in an annular radius of P-space lying between radii P and $P + dp$is

$$dN = \frac{3V}{\Lambda^3}p^2\,dp.$$

Hence, the total kinetic energy of the electron gas can be written

$$K = \frac{3V}{\Lambda^3}\int_0^{p_F}\frac{p^2}{2m}p^2 = \frac{3}{5}\frac{V}{\Lambda^3}\frac{p_F^5}{2m},$$

where m is the electron mass. It follows that

$$K = \frac{3}{5} N \frac{V^2}{2m} \left(\frac{N}{V} \right)^{2/3}.$$

The interior of a white-dwarf star is composed of atoms like C^{12} and C^{16} which contain equal numbers of protons, neutrons, and electrons. Thus,

$$M = 2\,N\,m_p,$$

where m_p is the proton mass.

Equations can be combined to give

$$E = \frac{A}{R^2} - \frac{B}{R},$$

where

$$A = \frac{3}{20} \left(\frac{9\pi}{8} \right)^{2/3} \frac{\hbar^2}{m} \left(\frac{M}{m_p} \right)^{5/3},$$

$$B = \frac{3}{5} G\,M^2.$$

The equilibrium radius of the star R_* is that which minimizes the total energy E. In fact, it is easily demonstrated that

$$R_* = \frac{2A}{B},$$

which yields

$$R_* = \frac{(9\pi)^{2/3} \hbar^2\, \hbar^2}{8} \frac{1}{m\,G\,m_p^{5/3} M^{2/3}}.$$

The above formula can also be written

$$\frac{R_*}{R_\odot} = 0.010 \left(\frac{M_\odot}{M} \right)^{1/3},$$

where $R_\odot = 7 \times 10^5$ kmis the solar radius, and $M_\odot = 2 \times 10^{30}$ kg is the solar mass. It follows that the radius of a typical solar mass white-dwarf is about 7000km: i.e., about the same as the radius of the Earth. The first white-dwarf to be discovered (in

1862) was the companion of Sirius. Nowadays, thousands of white-dwarfs have been observed, all with properties similar to those described above.

THE CHANDRASEKHAR LIMIT

One curious feature of white-dwarf stars is that their radius decreases as their mass increases. It follows, from Eq. that the mean energy of the degenerate electrons inside the star increases strongly as the stellar mass increases: in fact, $K \propto M^{4/3}$. Hence, if M becomes sufficiently large the electrons become relativistic, and the above analysis needs to be modified. Strictly speaking, the non-relativistic analysis is only valid in the low mass limit $M \ll M_{\Box}$. Let us, for the sake of simplicity, consider the ultra-relativistic limit in which $p \gg m\,c$.

The total electron energy (including the rest mass energy) can be written

$$K = \frac{3V}{\Lambda^3} \int_0^{PF} (p^2c^2 + m^2c^4)^{1/2} p^2 \, dp,$$

by analogy with Eq. Thus,

$$K \simeq \frac{3V\,c}{\Lambda^3} \int_0^{PF} \left(p^3 + \frac{m^2c^2}{2} p + \cdots \right) dp,$$

giving

$$E \simeq \frac{3\,V\,c}{4\,\Lambda^3} + \left[p_F^4 + m^2\,c^2 p_F^2 + \cdots \right].$$

It follows, from the above, that the total energy of an ultra-relativistic white-dwarf star can be written in the form

$$E \simeq \frac{A - B}{R} + C\,R,$$

where

$$A = \frac{3}{8} \left(\frac{9\pi}{8} \right)^{1/3} \hbar\, c \left(\frac{M}{m_p} \right)^{4/3},$$

$$B = \frac{3}{5} G\,M^2,$$

$$C = \frac{3}{4}\frac{1}{(9\pi)^{1/3}}\frac{m^2c^3}{\hbar}\left(\frac{M}{m_p}\right)^{2/3}.$$

As before, the equilibrium radius R_* is that which minimizes the total energy E. However, in the ultra-relativistic case, a non-zero value of R_* only exists for $A - B > 0$. When $A - B < 0$ the energy decreases monotonically with decreasing stellar radius: in other words, the degeneracy pressure of the electrons is incapable of halting the collapse of the star under gravity. The criterion which must be satisfied for a relativistic white-dwarf star to be maintained against gravity is that

$$\frac{A}{B} > 1.$$

This criterion can be re-written

$$M < M_C,$$

where

$$M_C = \frac{15}{64}(5\pi)^{1/2}\frac{(\hbar\, c/G)^{1/2}}{m_p^2} = 1.72\, M_\odot$$

is known as the Chandrasekhar limit, after A. Chandrasekhar who first derived it in 1931. A more realistic calculation, which does not assume constant density, yields

$$M_C = 1.4\, M_\odot.$$

Thus, if the stellar mass exceeds the Chandrasekhar limit then the star in question cannot become a white-dwarf when its nuclear fuel is exhausted, but, instead, must continue to collapse. What is the ultimate fate of such a star?

NEUTRON STARS

At stellar densities which greatly exceed white-dwarf densities, the extreme pressures cause electrons to combine with protons to form neutrons. Thus, any star which collapses to such an extent that its radius becomes significantly less than that characteristic of a white-dwarf is effectively transformed into a gas of neutrons. Eventually, the mean separation between the neutrons becomes comparable with their de Broglie wavelength. At this point, it is possible for the degeneracy pressure of the neutrons to halt the collapse of the star. A star which is maintained against gravity in this manner is called a neutron star.

Neutrons stars can be analyzed in a very similar manner to white-dwarf stars. In fact, the previous analysis can be simply modified by letting $m_p \to m_p/2$ and $m \to m_p$. Thus, we conclude that non-relativistic neutrons stars satisfy the mass-radius law:

$$\frac{R_*}{R_\odot} = 0.00011\left(\frac{M_\odot}{M}\right)^{1/3},$$

It follows that the radius of a typical solar mass neutron star is a mere 10km. In 1967 Antony Hewish and Jocelyn Bell discovered a class of compact radio sources, called pulsars, which emit extremely regular pulses of radio waves. Pulsars have subsequently been identified as rotating neutron stars. To date, many hundreds of these objects have been observed.

When relativistic effects are taken into account, it is found that there is a critical mass above which a neutron star cannot be maintained against gravity. According to our analysis, this critical mass, which is known as the Oppenheimer-Volkoff limit, is given by

$$M_{\odot V} = 4\, M_C = 6.9 M_\odot .$$

A more realistic calculation, which does not assume constant density, does not treat the neutrons as point particles, and takes general relativity into account, gives a somewhat lower value of

$$M_{\odot V} = 1.5 - 2.5\, M_\odot .$$

A star whose mass exceeds the Oppenheimer-Volkoff limit cannot be maintained against gravity by degeneracy pressure, and must ultimately collapse to form a black-hole.

Kinetic Theory of Gases

BERNOULLI'S PICTURE

Daniel Bernoulli, in 1738, was the first to understand air pressure from a molecular point of view. He drew a picture of a vertical cylinder, closed at the bottom, with a piston at the top, the piston having a weight on it, both piston and weight being supported by the air pressure inside the cylinder. He described what went on inside the cylinder as follows: "let the cavity contain very minute corpuscles, which are driven hither and thither with a very rapid motion; so that these corpuscles, when they strike against the piston and sustain it by their repeated impacts, form an elastic fluid which will expand of itself if the weight is removed or diminished..."Sad to report, his insight, although essentially correct, was not widely accepted.

Most scientists believed that the molecules in a gas stayed more or less in place, repelling each other from a distance, held somehow in the ether. Newton had shown that PV = constant followed if the repulsion were inverse-square. It was rejected by the president, Humphry Davy, who pointed out that equating temperature with motion, as Herapath did, implied that there would be an absolute zero of temperature, an idea Davy was reluctant to accept.

And it should be added that no-one had the slightest idea how big atoms and molecules were, although Avogadro had conjectured that equal volumes of different gases at the same temperature and pressure contained equal numbers of molecules—his famous number—neither he nor anyone else knew what that number was, only that it was pretty big.

MOLECULAR ENERGY AND PRESSURE

It is not difficult to extend Bernoulli's picture to a quantitative description, relating the gas pressure to the molecular velocities. As a warm up exercise, let us consider a single perfectly elastic particle, of mass m, bouncing rapidly back and forth at speed v inside a narrow cylinder of length L with a piston at one end, so all motion is along the same line.What is the force on the piston?

Obviously, the piston doesn't feel a smooth continuous force, but a series of equally spaced impacts. However, if the piston is much heavier than the particle, this will have the same effect as a smooth force over times long compared with the interval between impacts.

So what is the value of the equivalent smooth force? Using Newton's law in the form force = rate of change of momentum, we see that the particle's momentum changes by 2mv each time it hits the piston. The time between hits is 2L/v, so the frequency of hits is v/2L per second. This means that if there were no balancing force, by conservation of momentum the particle would cause the momentum of the piston to change by 2mv′v/2L units in each second. This is the rate of change of momentum, and so must be equal to the balancing force, which is therefore

F = mv2/L

We now generalize to the case of many particles bouncing around inside a rectangular box, of length L in the x-direction. The total force on the side of area A perpendicular to the x-direction is just a sum of single particle terms, the relevant velocity being the component of the velocity in the x-direction. The pressure is just the force per unit area, P = F/A. Of course, we don't know what the velocities of the particles are in an actual gas, but it turns out that we don't need the details. If we sum N contributions, one from each particle in the box, each contribution proportional to vx2 for that particle, the sum just gives us N times the average value of vx2. That is to say,

$$P = F / A = Nm\overline{v_x^2} / LA = nm\overline{v_x^2} / V$$

where there are N particles in a box of volume V. Next we note that the particles are equally likely to be moving in any direction, so the average value of vx2 must be the same as that of vy2 or vz2, and since v2 = vx2 + vy2 + vz2, it follows that

$$P = Nm\overline{v^2} / 3V.$$

This is a surprisingly simple result! The macroscopic pressure of a gas relates directly to the average kinetic energy per molecule. Of course, in the above we have not thought about possible complications caused by interactions between particles, but in fact for gases like air at room temperature these interactions are very small.

Furthermore, it is well established experimentally that most gases satisfy the Gas Law over a wide temperature range:

PV = nRT

for n moles of gas, that is, n = N/NA, with NA Avogadro's number and R the gas constant.

Introducing Boltzmann's constant k = R/NA, it is easy to check from our result for the pressure and the ideal gas law that the average molecular kinetic energy is proportional to the absolute temperature,

$$\overline{E_k} = \overline{\frac{1}{2}mv^2} = \frac{3}{2}kT.$$

Boltzmann's constant k = 1.38.10-23 joules/K.

MAXWELL FINDS THE VELOCITY DISTRIBUTION

By the 1850's, various difficulties with the existing theories of heat, such as the caloric theory, caused some rethinking, and people took another look at the kinetic theory of Bernoulli, but little real progress was made until Maxwell attacked the problem in 1859. Maxwell worked with Bernoulli's picture, that the atoms or molecules in a gas were perfectly elastic particles, obeying Newton's laws, bouncing off each other with straight-line trajectories in between collisions. (Actually, there is some inelasticity in the collisions with the sides—the bouncing molecule can excite or deexcite vibrations in the wall, this is how the gas and container come to thermal equilibrium).

Maxwell realized that it was completely hopeless to try to analyze this system using Newton's laws, even though it could be done in principle, there were far too many variables to begin writing down equations. On the other hand, a completely detailed description of how each molecule moved was not really needed anyway. What was needed was some understanding of how this microscopic picture connected with the macroscopic properties, which represented averages over huge numbers of molecules.

The relevant microscopic information is not knowledge of the position and velocity of every molecule at every instant of time, but just the distribution function, that is to say, what percentage of the molecules are in a certain part of the container, and what percentage have velocities within a certain range, at each instant of time. For a gas in thermal equilibrium, the distribution function is independent of time. Ignoring tiny corrections for gravity, the gas will be distributed uniformly in the container, so the only unknown is the velocity distribution function.

VELOCITY SPACE

What does a velocity distribution function look like? Suppose at some instant in time one particular molecule has velocity $\vec{v} = (v_x, v_y, v_z)$. We can record this information by constructing a three-dimensional velocity space, with axes v_x, v_y, v_z, and putting in a point P1 representing the molecule's velocity (the red arrow is of course $\vec{v}$):

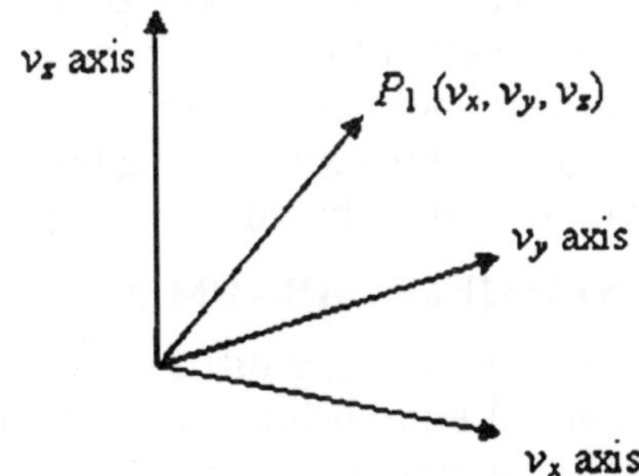

Fig. Point P_1 Represents the Location of one Molecule in Velocity Space

Now imagine that at that instant we could measure the velocities of all the molecules in a container, and put points P2, P3, P4, ... PN in the velocity space. Since N is of order 1021 for 100 ccs of gas, this is not very practical! But we can imagine what the result would be: a cloud of points in velocity space, equally spread in all directions (there's no reason molecules would prefer to be moving in the x-direction, say, rather than the y-direction) and thinning out on going away from the origin towards higher and higher velocities.

Now, if we could keep monitoring the situation as time passes individual points would move around, as molecules bounced off the walls, or each other, so you might think the cloud would shift around a bit. But there's a vast number of molecules in any realistic macroscopic situation, and for any reasonably sized container it's safe to assume that the number of molecules in any small region of velocity space remains pretty much constant. Obviously, this cannot be true for a region of velocity space so tiny that it only contains one or two molecules on average. But it can be shown statistically that if there are N molecules in a particular small volume of velocity space, the fluctuation of the number with time is of order $\sqrt{N}$, so a region containing a million molecules will vary in numbers by about one part in a thousand, a trillion molecule region by one part in a million. Since 100 ccs of air contains of order 1021 molecules, we can in practice divide the region of velocity space occupied by the gas into a billion cells, and still have variation in each cell of order one part in a million!

The bottom line is that for a macroscopic amount of gas, fluctuations in density, both in ordinary space and in position space, are for all practical purposes negligible, and we can take the gas to be smoothly distributed in both spaces.

MAXWELL'S SYMMETRY ARGUMENT

Maxwell found the velocity distribution function for gas molecules in thermal equilibrium by the following elegant argument based on symmetry.

For a gas of N particles, let the number of particles having velocity in the x-direction between vx and vx + dvx be $Nf_1(v_x)dv_x$. In other words, $Nf_1(v_x)dv_x$ is the fraction of all the particles having x-direction velocity lying in the interval between vx and vx + dvx. (I've written f1 instead of f to help remember this function refers to only one component of the velocity vector.)

If we add the fractions for all possible values of vx, the result must of course be 1:

$$\int_{-\infty}^{\infty} f_1(v_x)dv_x = 1.$$

But there's nothing special about the x-direction—for gas molecules in a container, at least away from the walls, all directions look the same, so the same function f will give the probability distributions in the other directions too. It follows immediately that the probability for the velocity to lie between vx and vx + dvx, vy and vy + dvy, and vz and vz + dvz must be:

$$Nf_1(v_x)dv_x f_1(v_y)dv_y f_1(v_z)dv_z = Nf_1(v_x)f_1(v_y)f_1(v_z)dv_x dv_y dv_z$$

Note that this distribution function, when integrated over all possible values of the three components of velocity, gives the total number of particles to be N, as it should (since integrating over each f1(v)dv gives unity).

Next comes the clever part—since any direction is as good as any other direction, the distribution function must depend only on the total speed of the particle, not on the separate velocity components. Therefore, Maxwell argued, it must be that:

$$f_1(v_x)f_1(v_y)f_1(v_z) = F\left(v_x^2 + v_y^2 + v_z^2\right).$$

where F is another unknown function. However, it is apparent that the product of the functions on the left is reflected in the sum of variables on the right. It will only come out that way if the variables appear in an exponent in the functions on the left. In fact, it is easy to check that this equation is solved by a function of the form:

$$f_1(v_x) = Ae^{-Bv_x^2}$$

This curve is called a Gaussian: it's centered at the origin, and falls off very rapidly as vx increases. Taking A = B = 1 just to see the shape, we find:

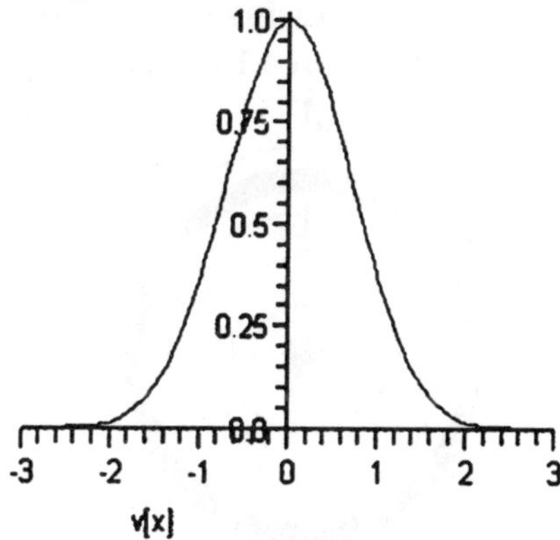

At this point, A and B are arbitrary constants—we shall eventually find their values for an actual sample of gas at a given temperature. Notice that (following Maxwell) we have put a minus sign in the exponent because there must eventually be fewer and fewer particles on going to higher speeds, certainly not a diverging number. Multiplying together the probability distributions for the three directions gives the distribution in terms of particle speed v, where v2 = vx2 +vy2 + vz2.

Since all velocity directions are equally likely, it is clear that the natural distribution function is that giving the number of particles having speed between v and v + dv.From the graph above, it is clear that the most likely value of vx is zero. If the gas molecules were restricted to one dimension, just moving back and forth on a line, then the most likely value of their speed would also be zero. However, for gas molecules free to move in two or three dimensions, the most likely value of the speed is not zero. It's easiest to see this in a two-dimensional example. Suppose we plot the points P representing the velocities of molecules in a region near the origin, so the density of points doesn't vary much over the extent of our plot.Now divide the two-dimensional space into regions corresponding to equal increments in speed:

$$0 \text{ to } \Delta v, \Delta v \text{ to } 2\Delta v, 2\Delta v \text{ to } 3\Delta v, \ldots$$

In the two-dimensional space, $v = \sqrt{v_x^2 + v_y^2}$ = constant is a circle, so this division of the plane is into annular regions between circles whose successive radii are Δv apart:

Constant speed circles in a two-dimensional example

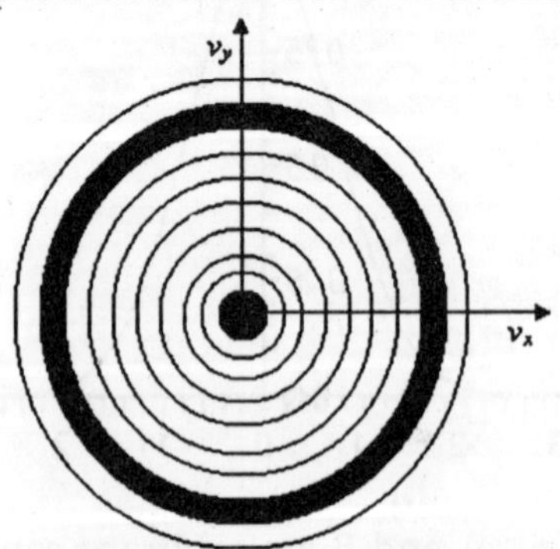

Fig. Constant Speed Circles in a Two-dimensional Example

Each of these annular areas corresponds to the same speed increment Δv. In particular, the green area, between a circle of radius $8\Delta v$ and one of radius $9\Delta v$, corresponds to the same speed increment as the small red circle in the middle, which corresponds to speeds between 0 and Δv. Therefore, if the molecular speeds are pretty evenly distributed in this near-the-origin area of the (vx, vy) plane, there will be a lot more molecules with speeds between $8\Delta v$ and $9\Delta v$ than between 0 and Δv—so the most likely speed will not be zero. To find out what it actually is, we have to put this area argument together with the Gaussian fall off in density on going far from the origin.

The same argument works in three dimensions—it's just a little more difficult to visualize. Instead of concentric circles, we have concentric spheres. All points lying on a spherical surface centered at the origin correspond to the same speed.

Let us now figure out the distribution of particles as a function of speed. The distribution in the three-dimensional space (v_x, v_y, v_z) is from Maxwell's analysis # of particles in small box

$$dv_x\, dv_y\, dv_z = Nf_1(v_x)f_1(v_y)f_1(v_z)dv_x dv_y dv_z$$
$$= NA^3 e^{-B\left(v_x^2+v_z^2+v_z^2\right)} dv_x dv_y dv_z$$
$$= NA^3 e^{-Bv^2} dv_x dv_y dv_z.$$

To translate this to the number of particles having speed between v and v + dv we need to figure out how many of those little $dv_x dv_y dv_z$ boxes there are corresponding to speeds between v and v + dv. In other words, what is the volume of velocity space between the two neighboring spheres, both centered at the origin, the inner one with radius v, the outer one infinitesimally bigger, with radius v + dv? Since dv is so tiny, this volume is just the area of the sphere multiplied by dv: that is, $4\pi v^2\, dv$.

Finally, then, the probability distribution as a function of speed is:

$$f(v)dv = 4\pi v^2 A^3 e^{-Bv^2} dv.$$

Of course, our job isn't over—we still have these two unknown constants A and B. However, just as for the function $f_1(v_x)$, $(v)dv$ is the fraction of the molecules corresponding to speeds between v and $v + dv$, and all these fractions taken together must add up to 1.

That is,

$$\int_0^\infty f(v)dv = 1.$$

We need the standard result $\int_0^\infty x^2 e^{-Bx^2} dx = (1/4B)\sqrt{\pi/B}$.

(a derivation can be found in my 152 Notes on Exponential Integrals), and find:

$$4\pi A^3 \frac{1}{4B}\sqrt{\frac{\pi}{B}} = 1.$$

This means that there is really only one arbitrary variable left: if we can find B, this equation gives us A: that is, $4\pi A^3 \frac{4}{\sqrt{\pi}} B^{3/2}$, and $4\pi A^3$ is what appears in $f(v)$.

Looking at $f(v)$, we notice that B is a measure of how far the distribution spreads from the origin: if B is small, the distribution drops off more slowly—the average particle is more energetic.

Recall now that the average kinetic energy of the particles is related to the temperature by $\overline{\frac{1}{2}mv^2} = \frac{3}{2}kT$. This means that B is related to the inverse temperature.

In fact, since $f(v)dv$ is the fraction of particles in the interval dv at v, and those particles have kinetic energy ½mv2, we can use the probability distribution to find the average kinetic energy per particle:

$$\overline{\tfrac{1}{2}mv^2} = \int_0^\infty \tfrac{1}{2}mv^2 f(v)dv$$

To do this integral we need another standard result:

$$\int_0^\infty x^4 e^{-Bx^2} dx = (3/8B^2)\sqrt{\pi/B}.$$

We find:

$$\overline{\tfrac{1}{2}mv^2} = \frac{3m}{4B}.$$

Substituting the value for the average kinetic energy in terms of the temperature of the gas,

$$\overline{\tfrac{1}{2}mv^2} = \tfrac{3}{2}kT$$

gives B = m/2kT, so $4\pi A^3 = \frac{4}{\sqrt{\pi}} B^{3/2} = 4\pi\left(\frac{m}{2\pi kT}\right)^{3/2}.$

This means the distribution function

$$f(v) = 4\pi\left(\frac{m}{2\pi kT}\right)^{3/2} v^2 e^{-mv^2/2kT} = 4\pi\left(\frac{m}{2\pi kT}\right)^{3/2} v^2 e^{-E/kT}$$

where E is the kinetic energy of the molecule.

Note that this function increases parabolically from zero for low speeds, reaches a maximum, then decreases exponentially. As the temperature increases, the position of the maximum shifts to the right. The total area under the curve is always one, by definition. For air molecules (say, nitrogen) at room temperature the curve is the blue one below. The red one is for an absolute temperature down by a factor of two:

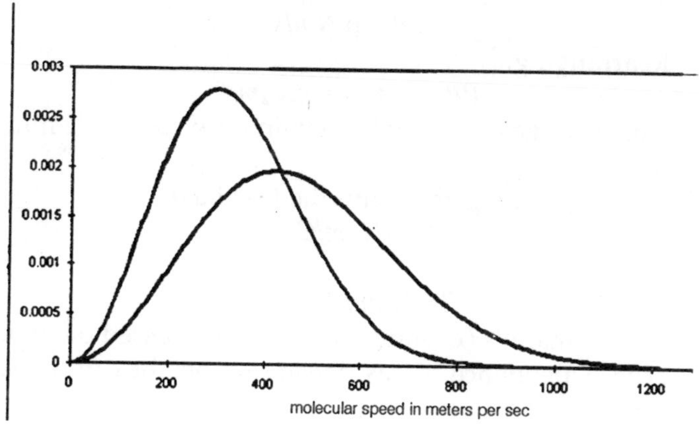

POTENTIAL ENERGY

Maxwell's analysis solves the problem of finding the statistical velocity distribution of molecules of an ideal gas in a box at a definite temperature T: the relative probability of a molecule having velocity $\vec{v}$ is proportional to $e^{-mv^2/2kT} = e^{-E/kT}$. The position distribution is taken to be uniform: the molecules are assumed to be equally likely to be anywhere in the box.

But how is this distribution affected if in fact there is some kind of potential pulling the molecules to one end of the box? In fact, we've already solved this problem, in the discussion earlier on the isothermal atmosphere. Consider a really big box, kilometers high, so air will be significantly denser towards the bottom. Assume the temperature is uniform throughout. We found under these conditions that with Boyles Law expressed in the form

$$\rho = CP$$

the atmospheric density varied with height as $P = P_0e^{-Cgh}$, or equivalently $\rho = \rho_0{}^{-Cgh}$.

Now we know that Boyle's Law is just the fixed temperature version of the Gas Law $PV = nRT$, and the density

$$\rho = \text{mass/volume} = Nm/V$$

with N the total number of molecules and m the molecular mass,

$$CP = \rho\, Nm/V.$$

Rearranging,

$$PV = Nm/C = nN_Am/C,$$

for n moles of gas, each mole containing Avogadro's number NA molecules.

Putting this together with the Gas Law,

$$PV = nN_Am/C = nRT,$$

so

$$C = N_Am/RT = m/kT$$

where Boltzmann's constant $k = R/N_A$ as discussed previously.

The dependence of gas density on height can therefore be written

$$\rho = \rho_0 e^{-Cgh} = \rho_0 e^{-mgh/kT}.$$

The important point here is that mgh is the potential energy of the molecule, and the distribution we have found is exactly parallel to Maxwell's velocity distribution, the potential energy now playing the role that kinetic energy played in that case.

We're now ready to put together Maxwell's velocity distribution with this height distribution, to find out how the molecules are distributed in the atmosphere, both in velocity space and in ordinary space. In other words, in a six-dimensional space!

Our result is:

$$f(x,y,z,v_x,v_y,v_z) = f(h,v) \propto e^{-mv^2/2kT} e^{-mgh/kT}$$

$$= e^{-((1/2)mv^2+mgh)/kT} = e^{-e/kT}$$

That is, the probability of a molecule having total energy E is proportional to $e^{-E/kT}$.

This is the Boltzmann, or Maxwell-Boltzmann, distribution. It turns out to be correct for any type of potential energy, including that arising from forces between the molecules themselves.

DEGREES OF FREEDOM AND EQUIPARTITION OF ENERGY

By a "degree of freedom" we mean a way in which a molecule is free to move, and thus have energy—in this case, just the x, y, and z directions. Boltzmann reformulated Maxwell's analysis in terms of degrees of freedom, stating that there was an average energy ½kT in each degree of freedom, to give total average kinetic energy 3.½kT, so the specific heat per molecule is presumable 1.5k, and given that k = R/NA, the specific heat per mole comes out at 1.5R. In fact, this is experimentally confirmed for monatomic gases. However, it is found that diatomic gases can have specific heats of 2.5R and even 3.5R.

This is not difficult to understand—these molecules have more degrees of freedom. A dumbbell molecule can rotate about two directions perpendicular to its axis. A diatomic molecule could also vibrate. Such a simple harmonic oscillator motion has both kinetic and potential energy, and it turns out to have total energy kT in thermal equilibrium. Thus, reasonable explanations for the specific heats of various gases can be concocted by assuming a contribution ½k from each degree of freedom. But there are problems. Why shouldn't the dumbbell rotate about its axis? Why do monatomic atoms not rotate at all? Even more ominously, the specific heat of hydrogen, 2.5R at room temperature, drops to 1.5R at lower temperatures. These problems were not resolved until the advent of quantum mechanics.

BROWNIAN MOTION

One of the most convincing demonstrations that gases really are made up of fast moving molecules is Brownian motion, the observed constant jiggling around of tiny particles, such as fragments of ash in smoke. This motion was first noticed by a Scottish botanist, who initially assumed he was looking at living creatures, but then found the same motion in what he knew to be particles of inorganic material. Einstein showed how to use Brownian motion to estimate the size of atoms.

IDEAL GAS THERMODYNAMICS:

The Kinetic Theory picture of a gas is often called the Ideal Gas Model. It ignores interactions between molecules, and the finite size of molecules. In fact, though, these only become important when the gas is very close to the temperature at which it become liquid, or under extremely high pressure. In this lecture, we will be analyzing the behavior of gases in the pressure and temperature range corresponding to heat engines, and in this range the Ideal Gas Model is an excellent approximation. Essentially, our program here is to learn how gases absorb heat and turn it into work, and vice versa. This heat-work interplay is called thermodynamics.

Julius Robert Mayer was the first to appreciate that there is an equivalence between heat and mechanical work. The tortuous path that led him to this conclusion is described in an earlier lecture, but once he was there, he realized that in fact the numerical equivalence—how many Joules in one calorie in present day terminology—could be figured out easily from the results of some measurements of gas specific heat by French scientists. The key was that they had measured specific heats both at constant volume and at constant pressure. Mayer realized that in the latter case, heating the gas necessarily increased its volume, and the gas therefore did work in pushing to expand its container. Having convinced himself that mechanical work and heat were equivalent, evidently the extra heat needed to raise the temperature of the gas at constant pressure was exactly the work the gas did on its container.

The simplest way to see what's going on is to imagine the gas in a cylinder, held in by a piston, carrying a fixed weight, able to move up and down the cylinder smoothly with negligible friction. The pressure on the gas is just the total weight pressing down divided by the area of the piston, and this total weight, of course, will not change as the piston moves slowly up or down: the gas is at constant pressure.

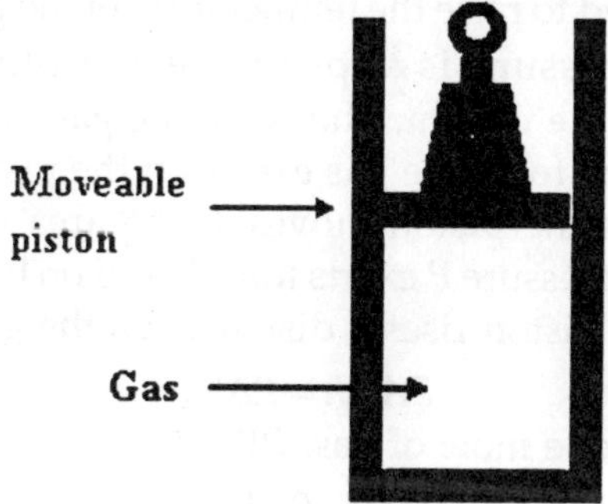

THE GAS SPECIFIC HEATS CV AND CP

Consider now the two specific heats of this same sample of gas, let's say one mole:

Specific heat at constant volume, C_V (piston glued in place),

Specific heat at constant pressure, C_P (piston free to rise, no friction).

In fact, we already worked out C_V in the Kinetic Theory lecture: at temperature T, recall the average kinetic energy per molecule is $\frac{3}{2}kT$, so one mole of gas—Avogadro's number of molecules—will have total kinetic energy, which we'll label internal energy,

$$E_{\text{int}}\,\tfrac{3}{2}kT.N_A\,\tfrac{3}{2}RT.$$

(In this simplest case, we are ignoring the possibility of the molecules having their own internal energy: they might be spinning or vibrating—we'll include that shortly).

That the internal energy is $\frac{3}{2}kT$ per mole immediately gives us the specific heat of a mole of gas in a fixed volume,

$$C_V = \tfrac{3}{2}R$$

that being the heat which must be supplied to raise the temperature by one degree.

However, if the gas, instead of being in a fixed box, is held in a cylinder at constant pressure, experiment confirms that more heat must be supplied to raise the gas temperature by one degree. As Mayer realized, the total heat energy that must be supplied to raise the temperature of the gas one degree at constant pressure is $\frac{3}{2}k$ per molecule plus the energy required to lift the weight. The work the gas must do to raise the weight is the force the gas exerts on the piston multiplied by the distance the piston moves. If the area of piston is A, then the gas at pressure P exerts force PA. If on heating through one degree the piston rises a distance Δh, the gas does work

$$PA.\Delta h = P\Delta V.$$

Now, for one mole of gas, $PV = RT$, so at constant P

$$P\Delta V = P\Delta T.$$

Therefore, the work done by the gas in raising the weight is just $R\Delta T$, the specific heat at constant pressure, the total heat energy needed to raise the temperature of one mole by one degree,

$$C_P = C_V + R.$$

In fact, this relationship is true whether or not the molecules have rotational or vibrational internal energy. (It's known as Mayer's relationship.) For example, the specific heat of oxygen at constant volume

$$C_V(O_2) = \tfrac{5}{2}R.$$

and this is understood as a contribution of $\frac{3}{2}R$ from kinetic energy, and R from the two rotational modes of a dumbbell molecule (just why there is no contribution form rotation about the third axis can only be understood using quantum mechanics). The specific heat of oxygen at constant pressure

$$C_P(O_2) = \frac{3}{2}R.$$

It's worth having a standard symbol for the ratio of the specific heats:

$$\frac{C_P}{C_V} = \gamma.$$

ISOTHERMS AND ADIABATS

An ideal gas in a box has three thermodynamic variables: P, V, T. But if there is a fixed mass of gas, fixing two of these variables fixes the third from $PV = nRT$ (for n moles). In a heat engine, heat can enter the gas, then leave at a different stage. The gas can expand doing work, or contract as work is done on it. To track what's going on as a gas engine transfers heat to work, say, we must follow the varying state of the gas. We do that by tracing a curve in the (P, V) plane. Supplying heat to a gas which consequently expands and docs mechanical work is the key to the heat engine. But just knowing that a gas is expanding and doing work is not enough information to follow its path in the (P, V) plane. The route it follows will depend on whether or not heat is being supplied (or taken away) at the same time. There are, however, two particular ways a gas can expand reversibly—meaning that a tiny change in the external conditions would be sufficient for the gas to retrace its path in the (P, V) plane backwards. It's important to concentrate on reversible paths, because as Carnot proved and we shall discuss later, they correspond to the most efficient engines. The two sets of reversible paths are the isotherms and the adiabats.

Isothermal behavior: The gas is kept at constant temperature by allowing heat flow back and forth with a very large object (a "heat reservoir") at temperature T. From $PV = nRT$, it is evident that for a fixed mass of gas, held at constant T but subject to (slowly) varying pressure, the variables P, V will trace a hyperbolic path in the (P, V) plane.

This path, $PV = nRT_1$, say is called the isotherm at temperature T1. Here are two examples of isotherms:

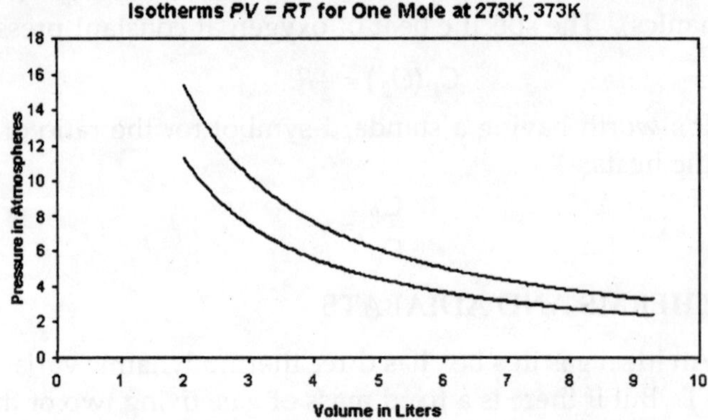

Adiabatic behavior: "adiabatic" means "nothing gets through", in this case no heat gets in or out of the gas through the walls. So all the work done in compressing the gas has to go into the internal energy E_{int}.

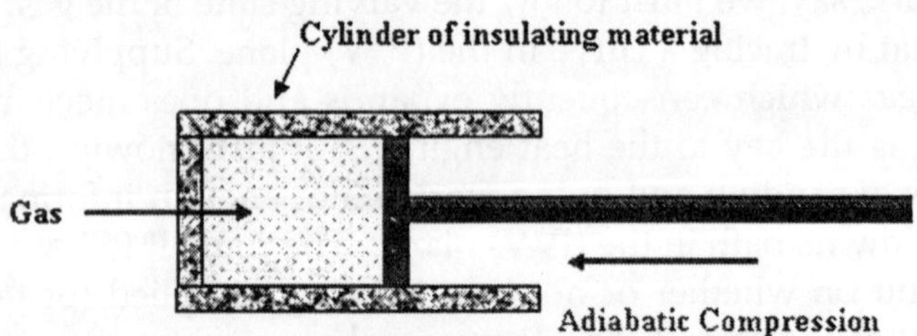

As the gas is compressed, it follows a curve in the (P, V) plane called an adiabat. To see how an adiabat differs from an isotherm, imagine beginning at some point on the blue 273K isotherm on the above graph, and applying pressure so the gas moves to higher pressure and lower volume.

Since the gas's internal energy is increasing, but the number of molecules is staying the same, its temperature is necessarily rising, it will move towards the red curve, then above it.

This means the adiabats are always steeper than the isotherms.

EQUATION FOR AN ADIABAT

What equation for an adiabat corresponds to $PV = nRT_1$ for an isotherm?

On raising the gas temperature by ΔT, the change in the internal energy—the sum of molecular kinetic energy, rotational energy and vibrational energy ,

$$\Delta E_{\text{int}} = C_V \Delta T.$$

This is always true: whether or not the gas is changing volume is irrelevant, all that counts in E_{int} is the sum of the energies of the individual molecules (assuming as we do here that attractive or repulsive forces between molecules are negligible). In adiabatic compression, all the work done by the external pressure goes into this internal energy, so

$$-P\Delta V = C_V \Delta T.$$

(Compressing the gas of course gives negative ΔV, positive ΔEint.)

To find the equation of an adiabat, we take the infinitesimal limit

$$-PdV = C_V dT.$$

Divide the left-hand side by PV, the right-hand side by RT (since $PV = RT$, that's OK) to find

$$-\frac{R}{C_V}\frac{dV}{V} = \frac{dT}{T}.$$

Recall now that $C_P = C_V + R$, and $C_P/C_V = \gamma$. It follows that

$$\frac{R}{C_V} = \frac{C_P - C_V}{C_V} = \gamma - 1.$$

Hence

$$-(\gamma - 1)\int\frac{dV}{V} = \int\frac{dT}{T}$$

and integrating

$$\ln T + (\gamma - 1)\ln V = \text{const.}$$

from which the equation of an adiabat is

$$TV^{-1} = \text{const}.$$

From $PV - RT$, the P, V equation for an adiabat can be found by multiplying the left-hand side of this equation by the constant PV/T, giving

$$PV^{\gamma} = \text{const. for an adiabat,}$$

where $\gamma = {}^5/_3$ for a monatomic gas, ${}^7/_5$ for a diatomic gas.

THE CARNOT CYCLE

All standard heat engines (steam, gasoline, diesel) work by supplying heat to a gas, the gas then expands in a cylinder and pushes a piston to do its work. The catch is that the heat and/or the gas must somehow then be dumped out of the cylinder to get ready for the next cycle.

We examine the first step, the expansion, then go on to the full cycle—Carnot's analysis. Carnot's aim was to figure out how to maximize the efficiency of a heat engine, and then work out what that efficiency was, that is, how much of the heat supplied was actually converted into the mechanical work done by the engine.

Remember that he had in mind the analogy of the water wheel, at that time still a main driving force of industry. He knew that the most efficient water wheels were those that operated smoothly, the water went into the buckets at the top from the same level, it didn't fall through any height, and didn't splash around. In the limit of a frictionless wheel, with gentle flow on and off the wheel, the machine would be reversible—turning it in reverse to raise the water back would take the same amount of work the wheel had delivered as the water fell.

This was clearly perfect efficiency, so these were to conditions to emulate in the heat engine. The analog to having the water flow into buckets at the same height, with no wasteful drop, is to have the heat from the heat supply flow into the gas at the same temperature. There must of course be a slight drop in temperature for the heat to flow at all, but this must be minimized. This means that as the heat is supplied and the gas expands, the temperature of the gas stays the same as that of the heat supply (the "heat reservoir") and the gas is expanding isothermally.

ISOTHERMAL EXPANSION

So the first question is: how much work is done by an isothermally expanding gas? Taking the temperature of the heat reservoir to be TH (H for hot), the expanding gas follows the isothermal path $PV = nRT_H$ in the (P, V) plane.

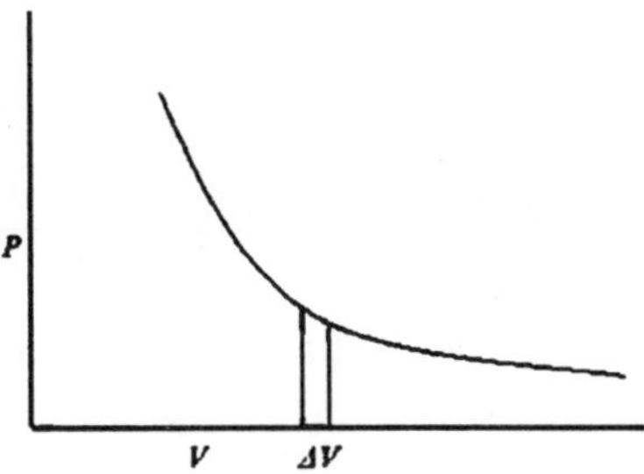

The work done by the gas in a small volume expansion ΔV is just $P\Delta V$, the area under the curve.Hence the work done in expanding isothermally from volume Va to Vb is the total area under the curve between those values,

$$\text{work done isothermally} = \int_{V_a}^{V_b} PdV = \int_{V_a}^{V_b} \frac{nRV_H}{V} dV = nRT_H \ln \frac{V_b}{V_a}.$$

Since the gas is at constant temperature TH, there is no change in its internal energy during this expansion, so the total heat supplied must be $nRT_H \ln \frac{V_b}{V_a}$, the same as the external work the gas has done.

In fact, this isothermal expansion is only the first step: the gas is at the temperature of the heat reservoir, hotter than its other surroundings, and will be able to continue expanding even if the heat supply is cut off. To ensure that this further expansion is also reversible, the gas must not be losing heat to the surroundings. That is, after the heat supply is cut off, there must be no further heat exchange with the surroundings, the expansion must be adiabatic.

ADIABATIC EXPANSION

The work done in an adiabatic expansion is like that done in allowing a compressed spring to expand against a force—

equal to the work needed to compress the spring in the first place, for a perfect spring, and an adiabatically enclosed gas is essentially perfect in this respect. In other words, adiabatic expansion is reversible.

To find the work the gas does in expanding adiabatically from Vb to Vc, say, the above analysis is repeated with the isotherm $PV = nRT_H$ replaced by the adiabat $PV^\gamma = P_b V_b^\gamma$, work done adiabatically W_{adiabat}

$$\int_{V_a}^{V_c} PdV = P_b V_b^\gamma \int_{V_b}^{V_c} \frac{dV}{V^\gamma} = P_b V_b^\gamma \frac{V_c^{1-\gamma} - V_b^{1-\gamma}}{1-\gamma}.$$

Again, this is the area under the curve, in this case under the adiabat, from b to c in the (P, V) plane.

Since points b, c are on the same adiabat, $P_b V_b^\gamma = P_c V_c^\gamma$, and the expression can be written more neatly:

$$W_{\text{adiabat}} = \frac{P_c V_c - P_b V_b}{1-\gamma}.$$

This is a useful expression for the work done since we are plotting in the (P, V) plane, but note that from the gas law $PV = nRT$, the numerator is just $nR(Tc - T_b)$, and from this $W_{\text{adiabat}} = nC_V(T_c - T_b)$, as of course it must be—this is the loss of internal energy that has been expended by the gas on expanding against external pressure.

We've looked in detail at the work a gas does in expanding as heat is supplied (isothermally) and when there is no heat exchange (adiabatically). These are the two initial steps in a heat engine, but it is equally necessary for the engine to get back to where it began, for the next cycle. The general idea is that the piston drives a wheel, which continues to turn and pushes the gas back to the original volume.

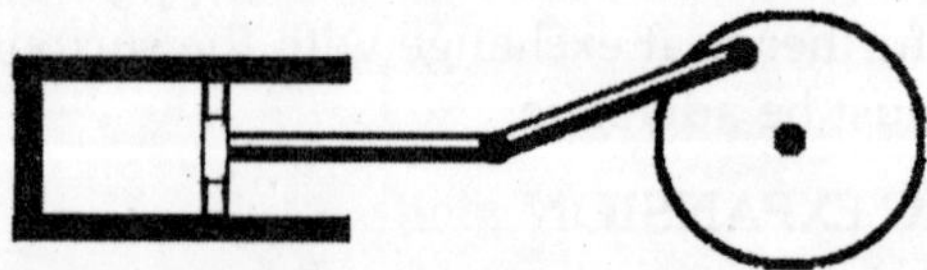

But it is also essential for the gas to be as cold as possible on this return leg, because the wheel is now having to expend work on the gas, and we want that to be as little work as possible—it's costing us. The colder the gas, the less pressure the wheel is pushing against.

To ensure that the engine is as efficient as possible, this return path to the starting point (P_a, V_a)must also be reversible. We can't just retrace the path taken in the first two legs, that would take all the work the engine did along those legs, and leave us with no net output. Now the gas cooled during the adiabatic expansion from b to c, from TH to TC, say, so we can go some distance back along the reversible colder isotherm TC. But this won't get us back to (P_a, V_a), because that's on the TH isotherm. The simplest option—the one chosen by Carnot—is to proceed back along the cold isotherm to the point where it intersects the adiabat through a, then follow that isotherm back to a. (One could follow a more complicated path: provided it was composed of segments each being adiabatic or isothermal, it would be reversible.) Carnot's cycle is around that curved quadrilateral having these four curves as its sides.

Efficiency of the Carnot Engine

In a complete cycle of Carnot's heat engine, the gas traces the path abcd. The important question is: what fraction of the heat supplied from the hot reservoir (along the red top isotherm) is turned into mechanical work? This fraction is called the efficiency of the engine.

The work output along any curve in the (P, V) plane is just $\int PdV$ —the area under the curve, but it will be negative if the volume is decreasing! So the work done by the engine during the hot isothermal segment is the area abfh, then the adiabatic expansion adds the area bcef, but as the gas is compressed back, the wheel has to do work on the gas equal to the area cdge as heat is dumped into the cold reservoir, then dahg as the gas is recompressed to the starting point.

The bottom line is that the total work done by the gas is the area bounded by the four paths: the curved "parallelogram" in the picture above. We could compute this area by finding $\int PdV$ for each segment, but that is unnecessary—on completing the cycle, the gas is back to its initial temperature, so has the same internal energy. Therefore, the work done by the engine must be just the difference between the heat supplied at TH and that dumped at TC.

Now the heat supplies along the initial hot isothermal path ab, equal to the work done along that leg, is (from the paragraph above on isothermal expansion):

$$Q_H = nRT_H \ln \frac{V_b}{V_a}$$

and the heat dumped into the cold reservoir along cd is

$$Q_C = nRT_C \ln \frac{V_c}{V_d}.$$

The difference between these two is the net work output. This can be simplified using the adiabatic equations for the other two sides of the cycle:

$$T_H V_b^{\gamma-1} = T_C V_c^{\gamma-1}$$

$$T_H V_a^{\gamma-1} = T_C V_d^{\gamma-1}.$$

Dividing the first of these equations by the second,

$$\left(\frac{V_b}{V_a}\right) = \left(\frac{V_c}{V_d}\right)$$

and using that in the preceding equation for QC,

$$Q_C = nRT_C \ln \frac{V_a}{V_b} = \frac{T_C}{T_H} Q_H.$$

The work done can now be written simply:

$$W = Q_H - Q_C = \left(1 - \frac{T_C}{T_H}\right) Q_H.$$

Therefore the efficiency of the engine, defined as the fraction of the ingoing heat energy that is converted to available work, is

$$\text{efficiency} = \frac{W}{Q_H} = 1 - \frac{T_C}{T_H}.$$

These temperatures are of course in degrees Kelvin, so for example the efficiency of a Carnot engine having a hot reservoir of boiling water and a cold reservoir ice cold water will be 1– (273/373) = 0.27, just over a quarter of the heat energy is transformed into useful work.After all the effort to construct an efficient heat engine, making it reversible to eliminate "friction" losses, etc., it is perhaps somewhat disappointing to find this figure of 27% efficiency when operating between 0 and 100 degrees Celsius. Surely we can do better than that? After all, the heat energy of hot water is the kinetic energy of the moving molecules, can't we find some device to channel all that energy into useful work? Well, we can do better than 27%, by having a colder cold reservoir, or a hotter hot one. But there's a limit: we can never reach 100% efficiency, because we cannot have a cold reservoir at $T_C = 0K$, and even if we did after the first cycle the heat dumped into it would warm it up!

The Second Law of Thermodynamics states that we cannot devise an engine, working in a cycle, that simply extracts heat from a hot reservoir and delivers mechanical work.

This means any engine that takes heat and delivers work also dumps out some of the initial heat to a reservoir at a lower temperature.

It's important to note that the First Law of Thermodynamics, the conservation of total energy including heat, would not be violated by an engine that powered a ship by extracting heat energy from the surrounding water. This Second Law is saying something new. And, this Second Law does not follow from the First by logical deduction—it comes (like the First) from experiment and observation.

THE MOST EFFICIENT ENGINE

An important consequence of the second Law is that no engine can be more efficient than the Carnot cycle. Essentially, this is because a "super efficient" engine, if one existed, could be used to drive a Carnot cycle in reverse, which would pump back to the hot reservoir the heat the super efficient engine dumped in the cold reservoir, and the net effect of the two coupled engines would be to take heat from the hot reservoir and do work, contradicting the Second Law.

To see this, we plot the heat/energy flow for the Carnot cycle:

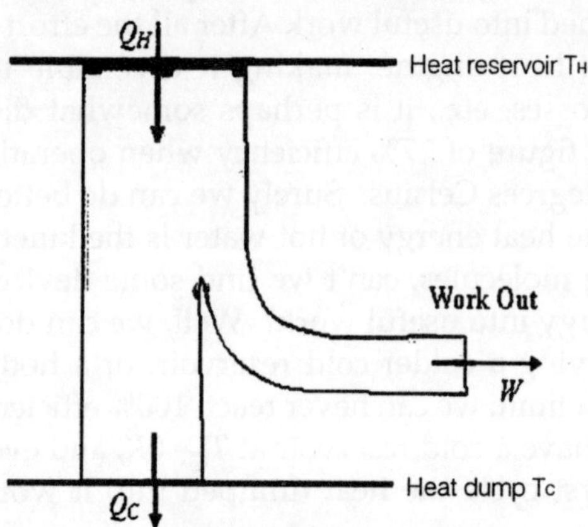

Here $Q_H = Q_C + W$ (all expressed in Joules, of course).

Since the engine is reversible, it can also be run backwards (this would be a refrigerator: outside work is supplied, and heat is extracted from a cold reservoir and dumped into a hot reservoir:

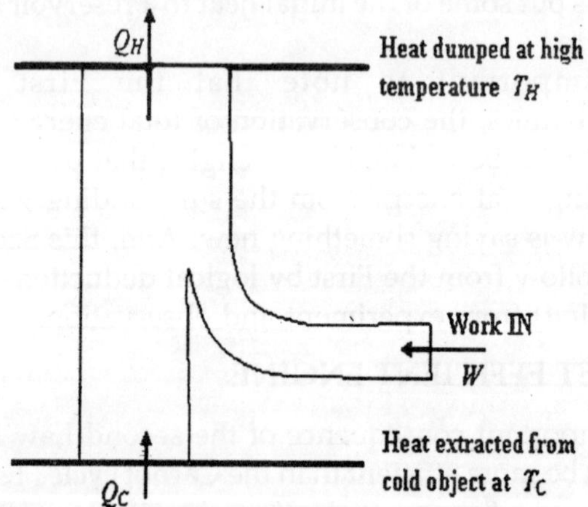

Suppose now we have a super efficient engine, represented by the first diagram above, and dumping the same heat per cycle QC into the cold reservoir, but taking in more heat energy $Q_H + \Delta$ Joules from the hot reservoir, and performing work $W + \Delta$. Now, we hook up our super efficient engine to the "Carnot refrigerator" in the diagram above. The refrigerator sucks out of the cold reservoir all the heat the super efficient engine dumped there, and needs W Joules of work per cycle to do it.

The super efficient engine can provide this, and there are still Δ Joules of work to spare. Of course, the Carnot refrigerator has also dumped QH Joules of heat in the hot reservoir. But the bottom line is that between them, the super efficient engine and the Carnot refrigerator have extracted Δ Joules of heat from the hot reservoir and performed Δ Joules of work—contradicting the Second Law.

The Second Law therefore forces the conclusion that no amount of machine design will produce an engine more efficient than the Carnot cycle. The rather low ultimate efficiencies this dictated came as a shock to nineteenth century engineers.

MOLECULAR COLLISIONS

In analyzing the gas so far, we've ignored collisions between molecules, and in fact for air at ordinary temperatures the relationship between pressure, volume and temperature came out correctly. Furthermore, Maxwell's speed distribution can be used to find what fraction of the molecules in a planet's atmosphere are moving at above escape velocity, so we can predict what gases will remain surrounding a planet, given the gravitational force near the surface, and the temperature. But there are other phenomena for which an understanding of collisions is all-important.

For example, if two different gases, say oxygen and nitrogen, in a container at the same pressure and temperature are separated by a partition, how quickly will they mix once the partition is removed? Assuming room temperature, the molecules will be moving at hundreds of meters per second, so one might imagine the mixing will be over in hundredths of a second. But that is not the case at all—observationally, it might take an hour, for a box holding a few liters. This surprisingly slow penetration of one gas by another is called diffusion. The reason it takes the gases so long to mix becomes evident on tracking one molecule as it enters the other gas. Think of an oxygen molecule moving into nitrogen. We'll take O2 and N2 to be little spheres of diameter d. So now visualize the little O2 sphere shooting into this space where all these other spheres are moving around. Temporarily, for ease of visualization, let's imagine all the other spheres to be at rest.

How far can we expect the O_2 to get before it hits an N_2? The average distance before a collision is called the mean free path. Let's try to picture how much room there is to fly between these fixed N_2 spheres. We do know that if it were liquid nitrogen, there would be very little room: liquids are just about incompressible, so the molecules must be touching. Roughly speaking, a molecule of diameter d will occupy a cubical volume of about d3 (there has to be some space left over—we can pack cubes to fill space, but not spheres).

We also know that liquid nitrogen weighs about 800 kg per cubic meter, whereas N_2 gas at room temperature (and pressure) weighs about 1.2 kg per cubic meter, a ratio of 670. This means that on average each molecule in the gas has 670 times more room—that is, it has a space 670 times the volume d3 we gave it in the liquid. So in the gas, the average center-to-center separation of the molecules will be the cube root of 670, which is about 8.75d. So the picture is a gas of spheres of diameter d, placed at random, but separated on average by distances of order 10d. It's clear that shooting an oxygen molecule into this it will get quite a way.

We now try to estimate just how far an O2 will get, on average, as it shoots into this forest of spheres. Picture the motion of the center of the oxygen molecule. Before any collision, it will be moving on a straight-line path. Just how close does the O_2 center have to get to an N2 center for a hit? Taking both O_2, N_2 to be spheres of diameter d, if an N2 center lies within d of the O_2 center's path, there will be a hit. So we can think of the O_2 as sweeping out a volume, a cylinder of radius d centered on its path, hitting and deflecting if it encounters an N_2 centered within that cylinder. So how far will it get, on average, before a hit? In traveling a distance x, it sweeps out a volume $\pi d^2 x$. Now picture it going through the gas for some considerable length of time, so there are many collisions.

The volume swept out will look like a stovepipe, long straight cylindrical sections connected by elbows at the collisions. The total volume of this stovepipe (ignoring tiny corrections from the elbows) will be just $\pi d^2 L$, L being the total length, that is, the total distance the molecule traveled. If the density of the nitrogen is n molecules per cubic meter, the number of N2's in this stovepipe volume will be $\pi d^2 Ln$, in other words, this will be the number of collisions. Therefore, the average distance between collisions, the mean free path l, is given by:

$$\text{mean free path } l = \frac{\text{total distance traveled}}{\text{number of collisions}} = \frac{L}{\pi d^2 Ln} = \frac{L}{\pi d^2 n}.$$

So what is n? We estimated above that each molecule has space 670d 3 to itself, so n is just how many of those volumes there are in one cubic meter, that is, $n = 1/670d^3$.

Therefore, the mean free path is given by

$$l = \frac{1}{\pi d^2 n} = \frac{670d^3}{\pi d^3} = 200d.$$

Notice that this derivation of the mean free path in terms of the molecular diameter depends only on knowing the ratio of the gas density to the liquid density—it does not depend on the actual size of the molecules! But it does mean that if we can somehow measure the mean free path, by measuring how fast one gas diffuses into another, for example, we can deduce the size of the molecules, and historically this was one of the first ways the size of molecules was determined, and so Avogadro's number was found.

Let us now put in some numbers to find this mean free path: for O2, N2, d = 0.3 nm, so the mean free path l = 60nm, or 6×10^{-8}m. The speed of the molecules at room temperature v is approximately 500 meters per sec., so the molecule has of order 1010 collisions per second!

Actually, there is one further correction we should make. We took the N2 molecules to be at rest, whereas in fact they're moving as fast as the oxygen molecule, approximately. This means that even if the O2 is temporarily at rest, it can undergo a collision as an N2 comes towards it. Clearly, what really counts in the collision rate is the relative velocity of the molecules. Defining the average velocity as the root mean square velocity, if the O2 has velocity $\vec{v}_1$ and the N2 $\vec{v}_2$, then the square of the relative velocity $\overline{(\vec{v}_1 - \vec{v}_2)^2} = \overline{\vec{v}_1^2} - 2\overline{\vec{v}_1 - \vec{v}_2}^2 + \overline{\vec{v}_2^2} = \overline{\vec{v}_1^2} + \overline{\vec{v}_2^2}$, since $\overline{\vec{v}_1.\vec{v}_2}$ must average to zero, the relative directions being random. So the average square of the relative velocity is twice the average square of the velocity, and therefore the average root-mean-square velocity is up by a factor "2, and the collision rate is increased by this factor. Consequently, the mean free path is decreased by a factor of "2 when we take into account that all the molecules are moving.

Our final result, then, is that the mean free path

$$1 = \frac{1}{\sqrt{2}\pi d^2 n}.$$

Finding the mean free path is—literally—the first step in figuring out how rapidly the oxygen atoms will diffuse into the nitrogen gas, and of course vice versa.

What we really want to know is just how much we can expect the gases to have intermingled after a given period of time. We'll just follow the one molecule, and estimate how far it gets. To begin with, let's assume for simplicity that it tales steps all of the same length l, but after each collision it bounces off in a random direction. So after N steps, it will have moved to a point

$$\vec{L} = \vec{l}_1 + \vec{l}_2 + \vec{l}_3 + \cdots + \vec{l}_N,$$

where each vector $\vec{l}_j$ has length l, but the vectors all point in random different directions.

If we now imagine many of the oxygen molecules following random paths like this, how far on average can we expect them to have drifted after N steps? (note that they could with equal likelihood be going backwards!) The appropriate measure is the root-mean-square distance,

$$\overline{\vec{L}^2} = \overline{\left(\vec{l}_1 + \vec{l}_2 + \vec{l}_3 ... \vec{l}_N\right)^2} = Nl^2 + \sum_{i.j} \overline{\vec{l}_i . \vec{l}_j}$$

Since the direction after each collision is completely random, $\overline{\vec{l}_i . \vec{l}_j} = 0$, and the root-mean-square distance $\sqrt{\overline{\vec{L}}^2} = \sqrt{N}l.$

If we allow steps of different lengths, the same argument works, but now l is the root-mean-square path length.

The important factor here is the $\sqrt{N}$. Recall from above that l = 60 nm, or 6×10^{-8} m., and there are of order 1010 collisions per second. This means that the average distance diffused in one second is $\sqrt{10^{10}}l = 10^5 l$, say half a centimeter. The average distance in one hour would be only 60 times this, or 30 cm., one foot, and in a day about five feet—the average distance traveled is only increasing as the square root of the time elapsed!

This is a very general result. For example, suppose we have a gas in which the mean free path is l and the average speed of the molecules is v. Then the average time between collisions $\tau = l/v$. The number of collisions in time t will be t/τ, so the average distance a molecule moves in time t will be $r = l\sqrt{t/\tau}$.

A famous mystery cleared up by arguments like this was that Newton predicted the speed of sound would be given by $c^2 = B/\rho$, as we discussed earlier in the course, with B the bulk modulus. But when B was measured carefully by slowly compressing air, the result was in error by about 30%! The speed of sound predicted a higher (stiffer) bulk modulus.

The explanation turned out to be that in a slow measurement of the bulk modulus, the gas stays at the same temperature—the heating caused by slow compression leaks away. But if the compression is rapid, the gas heats up and so the pressure goes up more than if it had stayed at the same temperature. So the question is whether the compression and decompression as a sound wave passes through is so rapid that the heated-up gas doesn't have time to spread to the cooled regions.

For sound at say 1000Hz, the wavelength is 34 cm. If compression heats gas locally, the hot molecules will diffuse away in a similar manner to that discussed above. They will be slightly faster than the average molecules. In 1/1000 th of a second, they will have 107 collisions, so will travel about $\sqrt{10^7}\, l$ = 30C $0l$ = 0.2mm. This tiny distance compared with the wavelength of the sound wave means that during the compression/decompression cycles as the wave passes through, the heat has no chance to dissipate—so, effectively, it's like compressing a gas in an insulated container, it's harder to compress than it would be if the heat generated could flow away, and the bulk modulus is higher by an amount (around 30%) we shall work out in a forthcoming lecture.

The simple kinetic picture of a gas leads to the Gas Law, $PV = nRT$, with the macroscopic variable T now seen as proportional to the average energy of a molecule. Putting heat into a gas simply means increasing the molecules' average kinetic energy (plus rotational energy, etc., for more complicated molecules) and the work done by an expanding gas is just a transfer of energy from this internal energy to whatever the gas is pushing.

The First Law of Thermodynamics is nothing but conservation of total energy, now that we know heat is really microscopic kinetic energy for an ideal gas (and rotational energy, etc.). By applying these findings to the Carnot Cycle, we discovered that this reversible ideal engine had efficiency $1 - T_C/T_H$. Furthermore, we found that no heat engine could be more efficient, at least if we granted the truth of the Second Law of Thermodynamics, which states that it is impossible to construct an engine that just takes heat energy out of the air, say, and converts it into work, without any available "cold reservoir" to dump waste heat into.

The incredible thing is that Carnot, in the 1820's, who believed heat was a conserved caloric fluid, and therefore

didn't know the First Law (that heat is just microscopic energy, which is not conserved by itself, but can be transformed into other forms of energy) correctly found the efficiency of his cycle to be $1 - T_C/T_H$, and then argued—also correctly—that this set the absolute limit on heat engine efficiency! How did he do that? It's worth a brief examination of his ideas, because they give a clue about something we haven't mentioned so far—entropy.

Remember, Carnot saw his cycle as a water wheel, the caloric fluid being the water, the temperature difference being the height difference from the top to the bottom of the waterfall. But he did know one difference from a waterfall: he knew there was an absolute zero, a level below which nothing could "fall". (This was from the well known results on the contraction of a gas with cooling. He took the absolute zero to be -267 Celsius, only a few degrees off the correct value). This meant that one could assign to the caloric fluid an absolute value of "potential energy" in this imaginary water wheel scenario: at temperature T, an amount of fluid F would have "potential energy" FT. This is the total possible amount of heat energy in the fluid, and to extract it all you would have to "drop" the fluid all the way down the temperature scale to the absolute zero. But in our real (if idealized) heat engine, it can only drop as far as the lowest temperature available, that of the cold reservoir TC. Hence the efficiency factor $1–T_C/T_H$: it's just the ratio of how far the caloric fluid is able to fall in our engine to how far it would have to fall (down from TH to absolute zero) to give up all its heat energy.

In this picture, the heat energy delivered from the hot reservoir $Q_H = FT_H$, so the amount of actual "caloric fluid" must be $F = Q_H/T_H$. But the fluid is conserved, in Carnot's picture, so the waste heat is $Q_C = FT_C$, with the same F, and therefore $Q_H/T_H = Q_C/T_C$. This equation is the important one: even though the argument he used is incorrect—there is no conserved caloric fluid—the equation is right, and relates directly to the efficiency of his engine: the heat utilized is $Q_H - Q_C = Q_H (1– T_C/T_H)$. But how did he know no engine could do better? Ironically, although he didn't know the First Law

of Thermodynamics, he had figured out the Second Law, and he presented an argument exactly like that given in the last lecture to demonstrate that no engine could outperform his reversible one.

Let's go back to the Carnot cycle and consider a bit more how the gas gains heat from a reservoir as it follows a path in the (P, V) plane. We begin with the first half of the cycle, from a to c:

ISOTHERM

This does not mean that we can say the gas at (P_C, V_C) has QH more heat than the gas at (P_a, V_a). Why not? Because we could equally well have gone from a to c by a route which is the second half of the Carnot cycle traveled backwards:

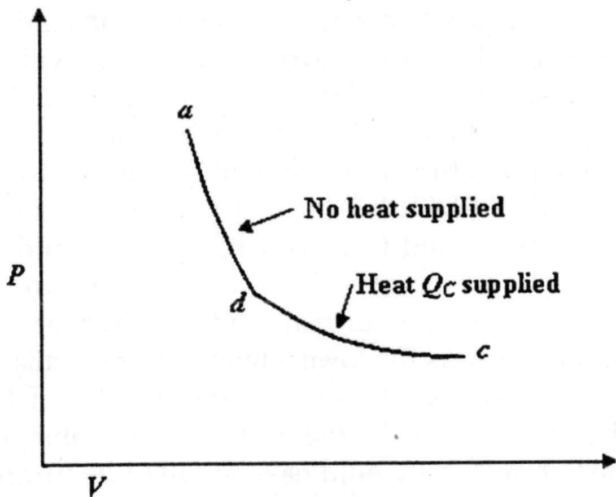

This is a perfectly well defined reversible route, ending at the same place, but with quite a different amount of heat supplied. So we cannot say that a gas at a given (P, V) contains a definite amount of heat. It does of course have a definite internal energy, but that energy can be increased by adding a mix of external work and supplied heat, and the two different routes from a to c have the same total energy supplied to the gas, but with more heat and less work along the top route.

However, notice that one thing is the same over the two routes in the diagrams above: the ratio of the heat supplied to the temperature at which is was delivered: $Q_H/T_H = Q_C/T_C$.

Of course, we've chosen two particular reversible routes from a to c, but it turns out that for any reversible route from a to c, the integral of increments of heat supplied divided by the temperature of delivery, $\int_a^c dQ/T$, is the same! To see how this can be, consider cutting a corner in the previous route:

If we follow the path aefgc instead of adc, where ef is an isotherm and fg an adiabat, how does that affect $\int_a^c dQ/T$? The answer is it doesn't: look at the little Carnot cycle efgd. We've just changed from the bottom route to the top route around this cycle from e to g, so from the previous argument about the original big Carnot cycle, $\int_e^g dQ/T$ is the same.

But we can now cut corners on the corners: any zigzag route from a to c, with the zigs isotherms and the zags adiabats, in other words, any reversible route, can be constructed by adding little Carnot cycles to the original route. In fact, any path you can draw in the plane from a to c can be approximated arbitrarily well by a reversible route made up of little bits of isotherms and adiabats.

The value of $\int_a^c dQ/T$ is evidently the same along any reversible path from a to c: in contrast to $\int_a^c dQ$, the heat change, which was different for different reversible routes.

We can define a new state variable, S, such that the difference

$$S(P_2, V_2) = S(P_1, V_1) + \int_{(P_1,V_1)}^{(P_2,V_2)} \frac{dQ}{T}$$

where the integral is understood to be along a reversible path. Given the value of S at a single point (P_1, V_1), we can find its value everywhere.

Chapter 8

Rotational Dynamics

The physical objects that we encounter in the world consist of collections of atoms that are bound together to form systems of particles. When forces are applied, the shape of the body may be stretched or compressed like a spring, or sheared like jello. In some systems the constituent particles are very loosely bound to each other as in fluids and gasses, and the distances between the constituent particles will vary. We shall begin our study of extended objects by restricting ourselves to an ideal category of objects, rigid bodies, which do not stretch, compress, or shear.

A body is called a rigid body if the distance between any two points in the body does not change in time. Rigid bodies, unlike point masses, can have forces applied at different points in the body. For most objects, treating as a rigid body is an idealization, but a very good one. In addition to forces applied at points, forces may be distributed over the entire body. Distributed forces are difficult to analyze; however, for example, we regularly experience the effect of the gravitational force on bodies. Based on our experience observing the effect of the gravitational force on rigid bodies, we note that the gravitational force can be concentrated at a point in the rigid body called the center of gravity, which for small bodies (so that $\vec{g}$ may be taken as constant within the body) is identical to the center of mass of the body.

Let's consider a rigid rod thrown in the air so that the rod is spinning as its center of mass moves with velocity $\vec{v}_{cm}$. Rigid bodies, unlike point-like objects, can have forces applied at different points in the body. We have explored the physics of translational motion; now, we wish to investigate the properties of rotation exhibited in the rod's motion, beginning with the notion that every particle is rotating about the center of mass with the same angular (rotational) velocity.

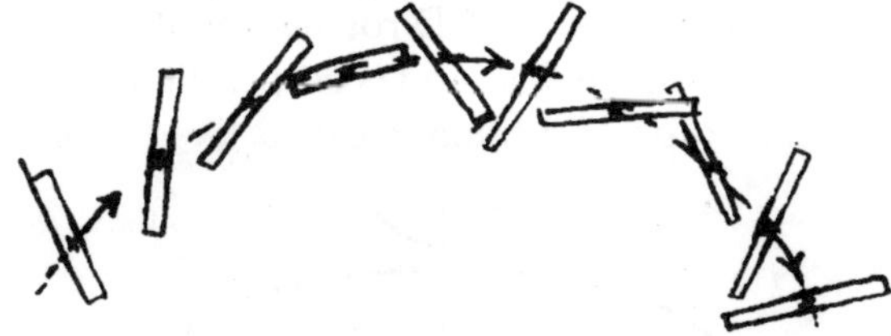

Fig. The Center of Mass of a thrown Rigid Rod follows a Parabolic Trajectory while the Rod Rotates about the Center of Mass.

We can use Newton's Second Law to predict how the center of mass will move. Since the only external force on the rod is the gravitational force (neglecting the action of air resistance), the center of mass of the body will move in a parabolic trajectory.

How was the rod induced to rotate? In order to spin the rod, we applied a torque with our fingers and wrist to one end of the rod as the rod was released. The applied torque is proportional to the angular acceleration. The constant of proportionality is called the moment of inertia. When external forces and torques are present, the motion of a rigid body can be extremely complicated while it is translating and rotating in space. We shall begin our study of rotating objects by considering the simplest example of rigid body motion, rotation about a fixed axis.

FIXED AXIS ROTATION: ROTATIONAL KINEMATICS

Fixed Axis Rotation

Static equilibrium, we demonstrated the need for two conditions: The total force acting on an object is zero, as is the total torque acting on the object. If the total torque is non-zero, then the object will start to rotate.

A simple example of rotation about a fixed axis is the motion of a compact disc in a CD player, which is driven by a motor inside the player. In a simplified model of this motion, the motor produces angular acceleration, causing the disc to spin. As the disc is set in motion, resistive forces oppose the motion until the disc no longer has any angular acceleration, and the disc now spins at a constant angular velocity. Throughout this process, the CD rotates about an axis passing through the center of the disc, and is perpendicular to the plane of the disc. This type of motion is called fixed-axis rotation.

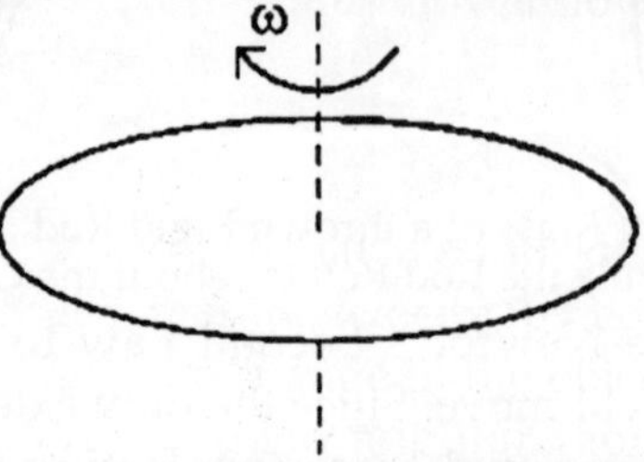

Fig. Rotation of a Compact Disc about a Fixed Axis.

When we ride a bicycle forward, the wheels rotate about an axis passing through the center of each wheel and perpendicular to the plane of the wheel. As long as the bicycle does not turn, this axis keeps pointing in the same direction. This motion is more complicated than our spinning CD because the wheel is both moving (translating) with some center of mass velocity, $\vec{v}_{cm}$, and rotating.

Fig. Fixed Axis Rotation and Center of Mass Translation for a Bicycle Wheel.

When we turn the bicycle's handlebars, we change the bike's trajectory and the axis of rotation of each wheel changes direction. Other examples of non-fixed axis rotation are the motion of a spinning top, or a gyroscope, or even the change in the direction of the earth's rotation axis.

This type of motion is much harder to analyze. Angular Velocity and Angular Acceleration. When we considered the rotational motion of a point-like object an angle coordinate è, and then defined the angular velocity as

$$\omega \equiv \frac{d\theta}{dt},$$

and angular acceleration (Equation 6.3.4) as

$$\alpha \equiv \frac{d^2\theta}{dt^2}.$$

For a rigid body undergoing fixed-axis rotation, we can divide the body up into small volume elements with mass Δm_i. Each of these volume elements is moving in a circle of radius r_1, i, about the axis of rotation

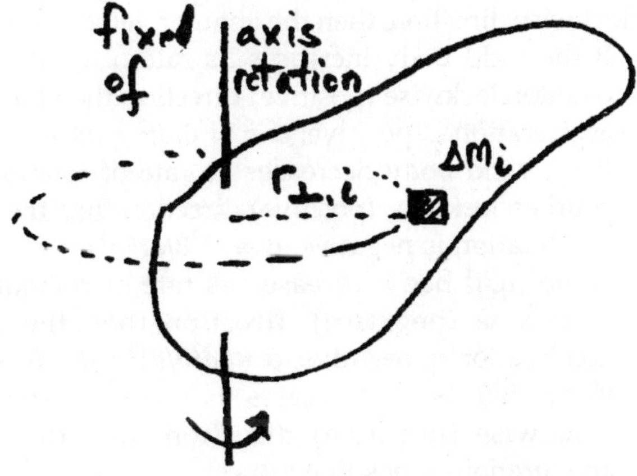

Fig. Coordinate System for Fixed-axis Rotation.

We will adopt the notation implied and denote the vector from the axis to the point where the mass element is located as $\vec{r}_{\perp i}$ with $\vec{r}_{\perp i} = |\vec{r}_{\perp i}|$,

Because the body is rigid, all the volume elements will have the same angular velocity ù and hence the same angular acceleration á. If the bodies did not have the same angular velocity, the volume elements would "catch up to" or "pass" each other, precluded by the rigid-body assumption.

ANGULAR VELOCITY AND ANGULAR ACCELERATION

Suppose we choose è to be increasing in the counterclockwise direction as

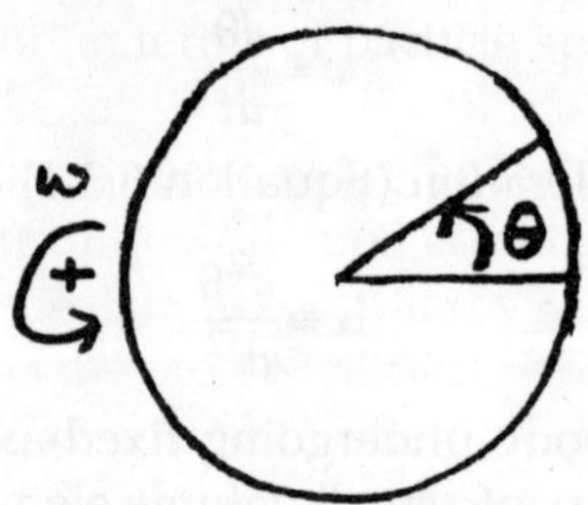

Fig. Sign Conventions for Rotational Motion.

If the rigid body rotates in the counterclockwise direction, then the angular velocity is positive, If the rigid body rotates in the clockwise direction, then the angular velocity is negative,

- If the rigid body increases its rate of rotation in the counterclockwise (positive) direction then the angular acceleration is positive, $\alpha \equiv$ º$d^2\theta/dt^2 = d\omega/dt > 0$.
- If the rigid body decreases its rate of rotation in the counterclockwise (positive) direction then the angular acceleration is negative, $\alpha \equiv$ º$d^2\theta/dt^2 = d\omega/dt > 0$.
- If the rigid body increases its rate of rotation in the clockwise (negative) direction then the angular acceleration is negative, $\alpha \equiv$ º$d^2\theta/dt^2 = d\omega/dt > 0$.
- If the rigid body decreases its rate of rotation in the clockwise (negative) direction then the angular acceleration is positive,

To phrase this more generally, if á and ù have the same sign, the body is speeding up; if opposite signs, the body is slowing down. This general result is independent of the choice of positive direction of rotation.

Tangential Velocity and Tangential Acceleration

Since the small volume Δm_i element of mass is moving in a circle of radius $r_{\perp i} = |\vec{r}_{\perp i}|$ with angular velocity ù, the element has a tangential velocity component

$$v_{\tan,i} = r_{\perp,i}\omega.$$

If the magnitude of the tangential velocity is changing, the volume element undergoes a tangential acceleration given by

$$a_{\tan,i} = r_{\perp,i}\alpha.$$

that the volume element is always accelerating inward with magnitude

$$\left|a_{\tan,i}\right| = \frac{v_{\tan,i}^2}{r_{\perp,i}} = r_{\perp,i}\,\omega^2.$$

TORQUE

A torque (τ) in physics, also called a moment (of force), is a pseudo-vector that measures the tendency of a force to rotate an object about some axis (center). The magnitude of a torque is defined as the product of a force and the length of the lever arm (radius). Just as a force is a push or a pull, a torque can be thought of as a twist.

The SI unit for torque is the newton meter (N m). In U.S. customary units, it is measured in foot pounds (ft·lbf) (also known as 'pound feet'). The symbol for torque is τ, the Greek letter tau. The concept of torque, also called moment or couple, originated with the studies of Archimedes on levers. The rotational analogues of force, mass, and acceleration are torque, moment of inertia, and angular acceleration, respectively.

Explanation

The force applied to a lever multiplied by its distance from the lever's fulcrum, the length of the lever arm, is its torque. A force of three newtons applied two meters from the fulcrum, for example, exerts the same torque as one newton applied six meters from the fulcrum.

This assumes the force is in a direction at right angles to the straight lever. The direction of the torque can be determined by using the right hand grip rule: curl the fingers of your right hand the direction of rotation and stick your thumb out so it is aligned with the axis of rotation. Your thumb points in the direction of the torque vector.

Mathematically, the torque on a particle (which has the position r in some reference frame) can be defined as the cross product:

$$\tau = r \times F$$

where

r is the particle's position vector relative to the fulcrum

F is the force acting on the particle.

The torque on a body determines the rate of change of its angular momentum,

$$\tau = \frac{dL}{dt}$$

where

L is the angular momentum vector

t stands for time.

As can be seen from either of these relationships, torque is a vector, which points along the axis of the rotation it would tend to cause.

Units

Torque has dimensions of force times distance and the SI unit of torque is the "newton meter" (N m). Even though the order of "newton" and "meter" are mathematically interchangeable, the BIPM (Bureau International des Poids et Mesures) specifies that the order should be N m not m N. N·m is also acceptable.

The joule, which is the SI unit for energy or work, is also defined as 1 N m, but this unit is not used for torque. Since energy can be thought of as the result of "force times distance", energy is always a scalar whereas torque is "force cross distance" and so is a (pseudo) vector-valued quantity. The dimensional equivalence of these units, of course, is not simply a coincidence: a torque of 1 N m applied through a full revolution will require an energy of exactly 2ð joules. Mathematically,

$$E = \tau\theta$$

where

E is the energy

τ is torque

θ is the angle moved, in radians.

Other non-SI units of torque include "pound-force-feet" or "foot-pounds-force" or "ounce-force-inches" or "meter-kilograms-force" or "kilogrammeter" (kgm).

EXTENDED UNITS IN RELATION WITH ROTATION ANGLES

As a consequence of the previous equation, if you introduce the radian (rad) as part of the dimensional units in the SI units system, the torque could be measured using "newton meters per radian" (N m/rad), or "joules per radian" (J/rad), while the energy needed and spent to perform the rotation would be measured simply in "newton meters" or "joules".

In the strict SI system, angles are not given any dimensional unit, because they do not designate physical quantities, despite the fact that they are measurable indirectly simply by dividing two distances (the arc length and the radius): one way to conciliate the two systems would be to say that arc lengths are not measures of distances (given they are not measured over a straight line, and a full circle rotation returns to the same position, i.e. a null distance).

So arc lengths should be measured in "radian meter" (rad·m), differently from straight segment lengths in "meters" (m). In such extended SI system, the perimeter of a circle whose radius is one meter, will be two pi rad·m, and not just two pi meters.

If you apply this measure to a rotating wheel in contact with a plane surface, the center of the wheel will move across a distance measured in meters with the same value, only if the contact is efficient and the wheel does not slide on it: this does not happen in practice, unless the surface of contact is constrained and is then not perfectly plane (and can resist to the horizontal linear forces applied to the irregularities of the pseudo-plane surface of movement and to the surface of the pseudo-circular rotating wheel); but then the system generates friction that loses some energy spent by the engine: this lost energy does not change the measurement of the torque or the total energy spent in the system but the effective distance that has been made by the center of the wheel.

The difference between the efficient energy spent by the engine and the energy produced in the linear movement is lost in friction and sliding, and this explains why, when applying the same non-null torque constantly to the wheel, so that the wheel moves at a constant speed according to the surface in contact, there may be no acceleration of the center of the wheel: in that case, the energy spent will be directly proportional to the distance made by the center of the wheel, and equal to the energy lost in the system by friction and sliding.

For this reason, when measuring the effective power produced by a rotating engine and the energy spent in the system to generate a movement, you will often need to take into account the angle of rotation, and then, adding the radian in the unit system is necessary as well as making a difference between the measurement of arcs (in radian meter) and the measurement of straight segment distances (in meters), as a way to effectively compute the efficiency of the mobile system and the capacity of a motor engine to convert between rotational power (in radian watt) and linear power (in watts): in a friction-free ideal system, the two measurements would have equal value, but this does not happen in practice, each conversion losing energy in friction (it's easier to limit all losses of energy caused by sliding, by introducing mechanical constraints of forms on the surfaces of contacts).

Depending on works, the extended units including radians as a fundamental dimension may or may not be used.

SPECIAL CASES AND OTHER FACTS

Moment Arm Formula

A very useful special case, often given as the definition of torque in fields other than physics, is as follows:

$$|\tau| = (\text{moment arm}) \cdot \text{force}$$

The construction of the "moment arm" along with the vectors r and F mentioned above. The problem with this definition is that it does not give the direction of the torque but only the magnitude, and hence it is difficult to use in three-dimensional cases.

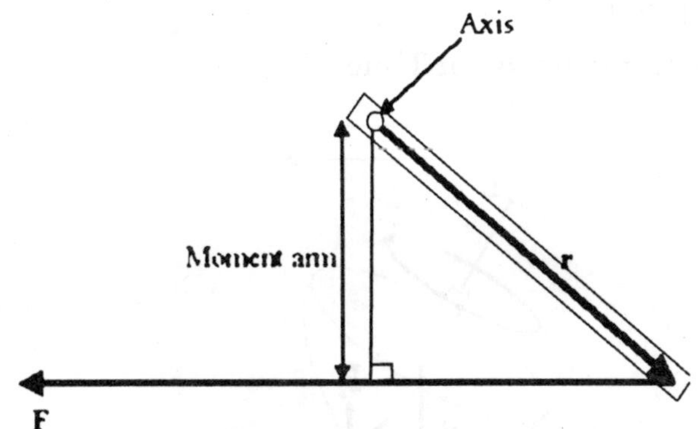

Fig. Moment Arm Diagram

If the force is perpendicular to the displacement vector r, the moment arm will be equal to the distance to the centre, and torque will be a maximum for the given force. The equation for the magnitude of a torque arising from a perpendicular force:

$$|\tau| = \text{(distance to center)} \cdot \text{force}$$

For example, if a person places a force of 10 N on a spanner (wrench) which is 0.5 m long, the torque will be 5 N m, assuming that the person pulls the spanner by applying force perpendicular to the spanner.

Force at an Angle

If a force of magnitude F is at an angle θ from the displacement arm of length r (and within the plane perpendicular to the rotation axis), then from the definition of cross product, the magnitude of the torque arising is:

$$\tau = rF\sin\theta$$

Static Equilibrium

For an object to be in static equilibrium, not only must the sum of the forces be zero, but also the sum of the torques (moments) about any point. For a two-dimensional situation with horizontal and vertical forces, the sum of the forces requirement is two equations: $\Sigma H = 0$ and $\Sigma V = 0$, and the torque a third equation: $\Sigma\tau = 0$. That is, to solve statically determinate equilibrium problems in two-dimensions, we use three equations.

Torque as a Function of Time

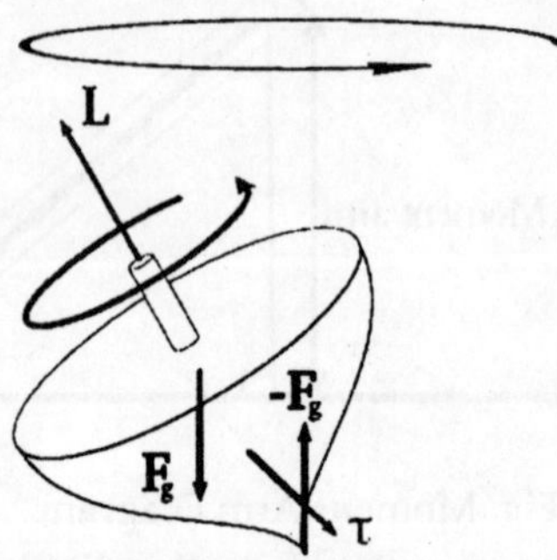

The torque caused by the two opposing forces Fg and -Fg causes a change in the angular momentum L in the direction of that torque. This causes the top to precess. Torque is the time-derivative of angular momentum, just as force is the time derivative of linear momentum:

$$\tau = \frac{dL}{dt}$$

where

L is angular momentum.

Angular momentum on a rigid body can be written in terms of its moment of inertia I and its angular velocity ω:

$$L = I\,\omega$$

so if I is constant,

$$\tau = I\frac{d\omega}{dt} I\alpha$$

where á is angular acceleration, a quantity usually measured in radians per second squared.

MACHINE TORQUE

Torque is part of the basic specification of an engine: the power output of an engine is expressed as its torque multiplied by its rotational speed of the axis. Internal-combustion engines produce useful torque only over a limited range of rotational speeds (typically from around 1,000–6,000 rpm for a small car). The varying torque output over that range can be measured with a dynamometer, and shown as a torque curve. The peak of that torque curve usually occurs somewhat below the overall power peak. The torque peak cannot, by definition, appear at higher rpm than the power peak.

Understanding the relationship between torque, power and engine speed is vital in automotive engineering, concerned as it is with transmitting power from the engine through the drive train to the wheels. Power is typically a function of torque and engine speed. The gearing of the drive train must be chosen appropriately to make the most of the motor's torque characteristics. Steam engines and electric motors tend to produce maximum torque close to zero rpm, with the torque diminishing as rotational speed rises (due to increasing friction and other constraints). Therefore, these types of engines usually have quite different types of drivetrains from internal combustion engines.

Torque is also the easiest way to explain mechanical advantage in just about every simple machine.

RELATIONSHIP BETWEEN TORQUE, POWER AND ENERGY

If a force is allowed to act through a distance, it is doing mechanical work. Similarly, if torque is allowed to act through a rotational distance, it is doing work. Power is the work per unit time. However, time and rotational distance are related by the angular speed where each revolution results in the circumference of the circle being travelled by the force that is generating the torque. The power injected by the applied torque may be calculated as:

$$\text{Power} = \text{torque} \cdot \text{angular speed}$$

On the right hand side, this is a scalar product of two vectors, giving a scalar on the left hand side of the equation. Mathematically, the equation may be rearranged to compute torque for a given power output. Note that the power injected by the torque depends only on the instantaneous angular speed

- not on whether the angular speed increases, decreases, or remains constant while the torque is being applied (this is equivalent to the linear case where the power injected by a force depends only on the instantaneous speed - not on the resulting acceleration, if any).

In practice, this relationship can be observed in power stations which are connected to a large electrical power grid. In such an arrangement, the generator's angular speed is fixed by the grid's frequency, and the power output of the plant is determined by the torque applied to the generator's axis of rotation.

Consistent units must be used. For metric SI units power is watts, torque is newton meters and angular speed is radians per second (not rpm and not revolutions per second).

Also, the unit newton meter is dimensionally equivalent to the joule, which is the unit of energy. However, in the case of torque, the unit is assigned to a vector, whereas for energy, it is assigned to a scalar.

Conversion to other Units

For different units of power, torque, or angular speed, a conversion factor must be inserted into the equation. Also, if rotational speed (revolutions per time) is used in place of angular speed (radians per time), a conversion factor of 2ð must be added because there are 2ð radians in a revolution:

$$\text{Power} = \text{torque} \times 2\pi \times \text{rotational speed},$$

where rotational speed is in revolutions per unit time.

Useful formula in SI units:

$$\text{Power(kW)} = \frac{\text{torque(N.m)} \times 2\pi \times \text{rotational speed (rpm)}}{60000}$$

where 60,000 comes from 60 seconds per minute times 1000 watts per kilowatt.

Some people (e.g. American automotive engineers) use horsepower (imperial mechanical) for power, foot-pounds (lbf·ft) for torque and rpm (revolutions per minute) for angular speed. This results in the formula changing to:

$$\text{Power(hp)} = \frac{\text{torque(lbf.ft)} \times 2\pi \times \text{rotational speed (rpm)}}{33000}$$

The constant below in, ft·lbf./min, changes with the definition of the horsepower; for example, using metric horsepower, it becomes ~32,550.

Use of other units (e.g. BTU/h for power) would require a different custom conversion factor.

Derivation

For a rotating object, the linear distance covered at the circumference in a radian of rotation is the product of the radius with the angular speed. That is: linear speed = radius x angular speed. By definition, linear distance=linear speed x time=radius x angular speed x time.

By the definition of torque: torque=force x radius. We can rearrange this to determine force=torque/radius. These two values can be substituted into the definition of power:

$$\text{Power} = \frac{\text{force} \times \text{linear distance}}{\text{time}} = \frac{\left(\frac{\text{torque}}{r}\right) \times (r \times \text{angular speed} \times t)}{t}$$

$$= \text{torque} \times \text{angular speed}$$

The radius r and time t have dropped out of the equation. However angular speed must be in radians, by the assumed direct relationship between linear speed and angular speed at the beginning of the derivation. If the rotational speed is measured in revolutions per unit of time, the linear speed and distance are increased proportionately by 2ð in the above derivation to give:

$$\text{Power} = \text{toqrque} \times 2\pi \times \text{rotational speed}$$

If torque is in lbf·ft and rotational speed in revolutions per minute, the above equation gives power in ft·lbf/min. The horsepower form of the equation is then derived by applying the conversion factor 33,000 ft·lbf/min per horsepower:

power = torque × 2π × rotational speed.

$$\frac{\text{ft.lbf}}{\text{min}} \times \frac{\text{horse power}}{33000.\frac{\text{ft.lbf}}{\text{min}}} \approx \frac{\text{torque} \times \text{RPM}}{5252}$$

because $5252.113122... = \dfrac{33,000}{2\pi}$.

MOMENT OF INERTIA

Newton's second law, Force = mass x acceleration, relates the acceleration that an object of a certain mass experiences when subject to a given force. There is an analogous relation between torque and angular acceleration, which introduces the concept of moment of inertia:

Torque = moment of inertia × angular acceleration

Just as mass is a measure of how readily an object accelerates due to a given force, the moment of inertia of an object measures how easily an object rotates about a particular point of rotation.

Thus, objects with a larger moment of inertia about a given point will be harder to rotate with a set torque. Correspondingly, a larger torque will cause a larger acceleration on a particular body.

The moment of inertia of a body, which is always measured relative to a point of rotation, depends in general on the object's mass and on its shape. It is perhaps evident that for a single mass going in a circle of fixed radius, the greater the radius the harder it is to change the angular velocity.

This is because the actual displacement, and hence linear velocity of the mass is proportional to the radius, so greater radius, for a given angular displacement means greater linear displacement.

In an extended object the parts that are further from the axis of rotation contribute more to the moment of inertia than the parts closer to the axis. So as a general rule, for two objects with the same total mass, the object with more of the mass located further from the axis will have a greater moment of inertia.

For example, the moment of inertia of a solid cylinder of mass M and radius R about a line passing through its center is MR2, whereas a hollow cylinder with the same mass and radius has a moment of inertia of MR2.

Similarly when a spinning figure skater pulls her arms in to her body she places more of her body weight closer to the axis of rotation and decreases her moment of inertia.

ROTATIONAL KINETIC ENERGY

Recall that an object of a certain mass moving with particular speed will have an associated kinetic energy$\frac{1}{2}$ mass x speed2. An object spinning about an axis will also have associated with it a kinetic energy, composed of the kinetic energies of each individual part of the object. These individual contributions may be summed up to give an expression for the total kinetic energy of the spinning object:

Kinetic energy = $\frac{1}{2}$ moment of inertia × (angular speed)^2

As with linear motion, where a force did work on an object and led to a change in the object's kinetic energy, for rotational motion the work done by a torque:

Work = torque × angular distance

goes into changing the rotational kinetic energy of an object:

Work done = change in kinetic energy

The rotational kinetic energy is treated like any other form of energy, in that it can be transformed into other forms (eg, potential), and also it is a component of the (conserved) total energy of a system.

Sometimes objects rotate about an axis that is itself in motion. For example, if you roll a cylinder down a ramp without any slipping, the axis about which it rotates (the center of the cylinder) moves down the ramp. In this case the total kinetic energy of the cylinder is the sum of its rotational kinetic energy plus its translational kinetic energy. It is therefore easy to see that an object with a higher moment of inertia will take longer to roll down the ramp. As a specific case consider a solid cylinder and a hollow cylinder with the same mass and radius starting at rest and rolling down the same ramp side by side. Which do you expect will reach the bottom first? We know that the moment of inertia of the hollow cylinder is greater, so that for a given angular velocity it has more rotational kinetic energy.

Moreover, if they both roll without slipping the linear velocity (and hence linear kinetic energy) is the same for both cylinders when their angular velocity is the same. In addition, the change in gravitational potential energy is the same for both cylinders (the weight of the cylinder times the height of the ramp), so both cylinders must have the same total kinetic energy at the bottom of the ramp.

This necessarily implies that the angular velocity (and linear velocity) of the hollow cylinder must be less at the bottom of the ramp than that of solid cylinder (otherwise it would have more kinetic energy). Consequently we can conclude that the hollow cylinder moves slower and the solid cylinder arrives at the bottom first.

ANGULAR MOMENTUM

Recall for linear motion that we introduced the concept of (linear) momentum, mass x velocity, by equating the force exerted on an object to the change in the object's momentum. An analogous quantity called angular momentum can be introduced in the same way. For an object rotating about an axis, the angular momentum is defined as

Angular momentum = moment of inertia × angular velcity

which leads to the net torque exerted on the object being expressed as

Torque = moment of inertia × angular acceleration
= change in angular momentum

Conservation of linear momentum is a powerful tool in analyzing, for example, collisions between objects undergoing linear motion.

Similarly, conservation of angular momentum is very useful in many situations. Consider again the ice skater spinning around. If she draws her arms inwards, she decreases her moment of inertia. Conservation of angular momentum requires that the skater's moment of inertia times her angular velocity remain constant.

Thus if her moment of inertia decreases, her angular velocity must increase, and this is what we observe: when the skater pulls in her arms, she immediately starts to rotate faster.

It is also worth noting that just like linear momentum, conservation of angular momentum is associated with a symmetry of the laws of physics. In this case, the relevant symmetry has to do with rotations about any axis.

THE PARALLEL AXIS THEOREM

The swingweight of a racquet is measured by the Babolat RDC using an axis of rotation 10 cm from the butt. To find the swingweight about the axis used on the stroke requires application of the Parallel Axis Theorem.

Swingweight, also known as moment of inertia and rotational inertia, is the resistance to change in the speed of the rotation about the axis of rotation. High swingweight means that the racquet is hard to get rotating, but once it gets going it will not be pushed around so much on impact with the ball and will tend to produce better pace and spin. Swingweight is the infinite sum of all infinitely small mass elements times the square of their distance from the axis of rotation. Or in mathematical terms,

$$I = \int r^2 dm$$

Call the known swingweight I and the unknown swingweight (about the axis of rotation used in the stroke) will be called I′. The swingweight of the racquet about its mass center will be called Ic. The distance of the mass center (balance point of the racquet) from the axis of rotation for the known swingweight is r, and the racquet mass is M. The distance of the mass center from the new axis of rotation is r + x. The Parallel Axis Theorem (from first year physics) tells us that the swingweight is the sum of the swingweight about the mass center (Ic) plus the product of the mass (M) and the square of the distance (r) from the axis of rotation to the mass center.

$$I = Ic + Mr2$$

So once we know the RDC swingweight about the 10 cm axis, we can find the swingweight about any other parallel axis by another application of the Parallel Axis Theorem. The first step is finding the swingweight about an axis through the mass center (Ic).

$$I = Ic + Mr2$$

$$\Rightarrow Ic = I - Mr2$$

The same holds for the unknown swingweight, which is about a parallel axis that is a distance r + x from the mass center:

$$I' = Ic + M(r + x)2$$

Now substitute what we've already found is equal to Ic:

$$I' = (I - Mr2) + M(r + x)2$$
$$= (I - Mr2) + M(r2 + 2rx + x2)$$

Simplifying, we get a general formula for finding the unknown swingweight (I') about a different axis (r + x):

$$I' = I + M(2rx + x2)$$

For the First Benchmark Condition (groundstroke), the axis of rotation for the stroke is at 7 cm from the handle end, so x is 3. And in the Second Benchmark Condition (serve), the axis is even farther away, and x is 5. The variable r is the published balance point minus 10 cm, but in our formulas we use r, which is the published balance point minus 7 or 5 cm according to which benchmark condition we are using, so we need to substitute in the above formula so we can use r instead of r. We will use another variable (a) to denote the axis used. So:

$$x = 10 - a$$

$$r = r + (10 - a) \Rightarrow r = r - (10 - a)$$

When we substitute in the above formula, we get:

$$I' = I + M(2rx + x2) = I + M[2(r - (10 - a))(10 - a) + (10 - a)2]$$

Now simplifying the expression in the brackets:

$$[2(r - (10 - a))(10 - a) + (10 - a)2]$$
$$= -2ar + 20a - 2a2 + 20r - 200 + 20a + 100 - 20a + a2$$
$$= -2ar + 20a - a2 + 20r - 100$$
$$= 20r - 2ar - 100 + 20a - a2$$
$$= 2r(10 - a) - (10 - a)2$$

So in the formulas, the variable I, which represents the swingweight about the axis used on the stroke, where a is the distance from the butt to the axis, I10 is the RDC swingweight about the axis 10 cm from the butt, and r is the distance in cm from the mass center (balance point) to the axis used in the stroke:

$$I = Ia = I10 + M*[2*(10 - a)*r - (10 - a)^2]$$

THE SIMPLE PENDULUM

Consider a mass suspended from a light inextensible string of length , such that the mass is free to swing from side to side in a vertical plane. This setup is known as a simple pendulum. Let be the angle subtended between the string and the downward vertical. Obviously, the equilibrium state of the simple pendulum corresponds to the situation in which the mass is stationary and hanging vertically down (i.e.,). The angular equation of motion of the pendulum is simply

$$I\ddot{\theta} = \tau,$$

where is the moment of inertia of the mass, and is the torque acting on the system. For the case in hand, given that the mass is essentially a point particle, and is situated a distance from the axis of rotation (i.e., the pivot point), it is easily seen that $I = ml^2$.

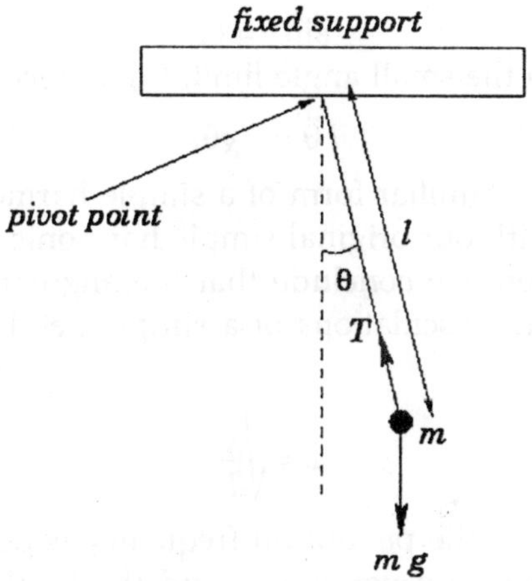

Fig. A Simple Pendulum.

The two forces acting on the mass are the downward gravitational force, *mg*, and the tension, *T*, in the string. Note, however, that the tension makes no contribution to the torque, since its line of action clearly passes through the pivot point. From simple trigonometry, the line of action of the gravitational force passes a distance from the pivot point. Hence, the magnitude of the gravitational torque is $mg\,l \sin\theta$. Moreover, the gravitational torque is a restoring torque: i.e., if the mass is displaced slightly from its equilibrium state (i.e., $\theta = 0$) then the gravitational force clearly acts to push the mass back toward that state. Thus, we can write

$$\tau = -mgl\sin\theta.$$

Combining the previous two equations, we obtain the following angular equation of motion of the pendulum:

$$l\ddot{\theta} = -g\sin\theta.$$

Unfortunately, this is not the simple harmonic equation. Indeed, the above equation possesses no closed solution which can be expressed in terms of simple functions.

Suppose that we restrict our attention to relatively small deviations from the equilibrium state. In other words, suppose that the angle θ is constrained to take fairly small values. We know, from trigonometry, that for $|\theta|$ less than about 6° it is a good approximation to write

$$\sin\theta \simeq \theta.$$

Hence, in the small angle limit, Eq. reduces to

$$l\ddot{\theta} = -g\theta,$$

which is in the familiar form of a simple harmonic equation. Comparing with our original simple harmonic equation, Eq. and its solution, we conclude that the angular frequency of small amplitude oscillations of a simple pendulum is given by

$$\omega = \sqrt{\frac{g}{l}}.$$

In this case, the pendulum frequency is dependent only on the length of the pendulum and the local gravitational acceleration, and is independent of the mass of the pendulum and the amplitude of the pendulum swings (provided that $\sin\theta \simeq \theta$ remains a good approximation). Historically, the simple pendulum was the basis of virtually all accurate time-keeping devices before the advent of electronic clocks. Simple pendulums can also be used to measure local variations ing.

Index

V

Z